EXPLORING THE NIGHT SKY

THE EQUINOX ASTRONOMY GUIDE FOR BEGINNERS

TERENCE DICKINSON

Principal Illustrations
by John Bianchi

FIREFLY BOOKS

First published in 1987 by Camden House Publishing
(a division of Telemedia Communications Inc.)

Twelfth printing 1998

Cataloguing-in-Publication Data

Dickinson, Terence
Exploring the night day

Includes index.
ISBN 0-920656-64-1 (bound)
ISBN 0-920656-66-8 (pbk.)

1. Astronomy – Juvenile literature. I. Title.

QB46.D52 1987 j520 C87-093567-4

Published by
Firefly Books Ltd.
3680 Victoria Park Avenue
Willowdale, Ontario
Canada M2H 3K1

Published in the U.S. by
Firefly Books (U.S.) Inc.
P.O. Box 1338, Ellicott Station
Buffalo, New York 14205

To my nephews, Christopher and Andrew,
and to questioning young minds everywhere.

Design by
Ulrike Bender

Front Cover: Night sky montage.
Illustration by John Bianchi.

Back Cover: Comet nucleus in deep space.
Illustration by John Bianchi.

Color separations by
Herzig Somerville Limited, Toronto, Ontario

Printed and bound in Canada by
Friesens
Altona, Manitoba

Printed on acid-free paper

Other books by Terence Dickinson:
NightWatch
The Universe and Beyond
Exploring the Sky by Day
The Backyard Astronomer's Guide (with Alan Dyer)
From the Big Bang to Planet X
Extraterrestrials
Other Worlds
Summer Stargazing
Splendors of the Universe (with Jack Newton)

CONTENTS

A COSMIC VOYAGE

ALIEN VISTAS

STARGAZING

OUR UNIVERSE: FROM EARTH TO THE DISTANT GALAXIES

The starry night sky has intrigued humans for thousands of years. To our ancestors, the stars and planets were mysterious lights in the darkness above. They could only guess about what was really out there. Gradually, over the centuries, astronomers began to understand what they were seeing. Today, we know the universe contains more galaxies than there are people on Earth. And each galaxy has as many stars as there are grains of sand in an overflowing wheelbarrow. Some of the stars are so enormous that it would take years for a spacecraft just to get from one side to the other. Some are no wider than a small city. Our sun is about midway between these extremes.

The sun is a star still in its youth. Stars do not exist forever. Like people, they are born and they die. Astronomers have gathered evidence that the entire universe has not existed forever either. About 15 billion years ago, a majestic explosion called the Big Bang created an expanding bubble of energy. After millions of years, the bubble cooled into gas. Galaxies then began to form from the gas. Later, stars were born within the galaxies. One of those stars was the sun. Astronomers think the sun was born 4½ billion years ago. Earth and the other planets in the solar system were formed along with the sun from leftover material.

That's the big picture, but there are lots of details to be filled in. This book is divided into three sections. The first, a 10-step voyage from the Earth's vicinity to the distant reaches of the universe, sets the stage for part two, "Alien Vistas," a sequence of 10 close-up looks at some of the most interesting objects mentioned in the first section. The final segment is a guide to viewing the night sky, which will enable you to go outside on any clear night of the year and identify celestial objects. Use the glossary at the back of the book for explanations of unfamiliar terms and for pronunciations.

Spaceship Earth, our home, above. Of the solar system's nine planets and four dozen moons, it is the only world that has liquid water. North America is the brown landmass seen near the centre of this photograph taken by the Apollo astronauts.

The known universe contains more than a billion trillion stars. The photograph at right shows a few thousand of them within the Milky Way Galaxy. The pink wisps are the gaseous remains of a star that exploded about 10,000 years ago.

STEP ONE: 1.3 LIGHT-SECONDS FROM EARTH

Nothing in our universe moves faster than light. Laser beams shot from Earth reach the moon in less than 1½ seconds. Compared with light-speed, our fastest space probes barely crawl along. They take hours to get to the moon. For a spacecraft to travel to planets such as Mars or Jupiter, a voyage of months or years is required.

The time it takes for light to go from one place to another in space is a convenient way of comparing distances. A light-second is the distance light travels in one second, about 186,000 miles (300,000 km). The moon is 1.3 light-seconds from Earth. Our planet is 500 light-seconds, or 8.3 light-minutes, from the sun. Therefore, sunlight that is shining on Earth right now left the sun's surface 8.3 minutes ago.

The moon, our closest celestial neighbour, is one-quarter the Earth's diameter – like an orange compared with a soccer ball. If a spaceship from another solar system were to approach Earth, the alien science officer might record in the ship's log that the third planet from the sun has a satellite as large as a small planet. The science officer would have seen places like the moon – airless and cratered, unchanging and lifeless – many times before. But Earth would attract attention because its surface is almost entirely covered with water. Worlds like Earth are probably rare. Just how rare nobody can guess. We simply do not know enough about how planets are born. Some stars may have families of planets, but many could have none.

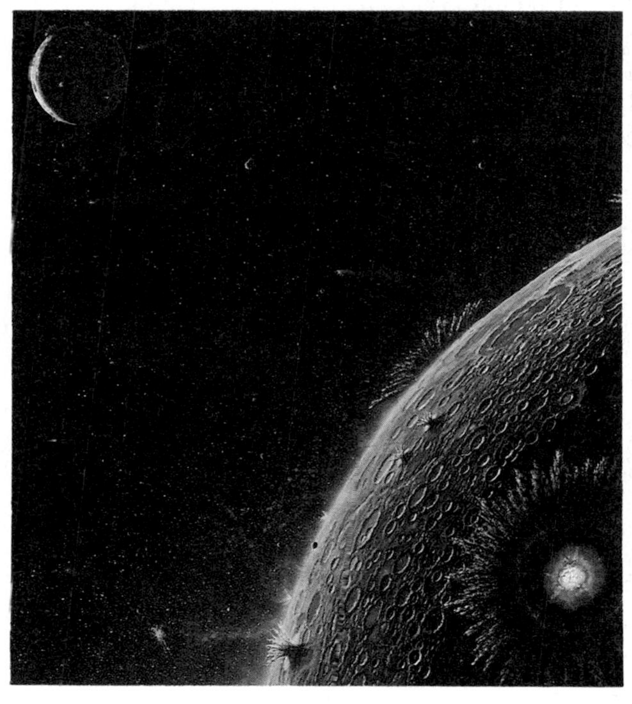

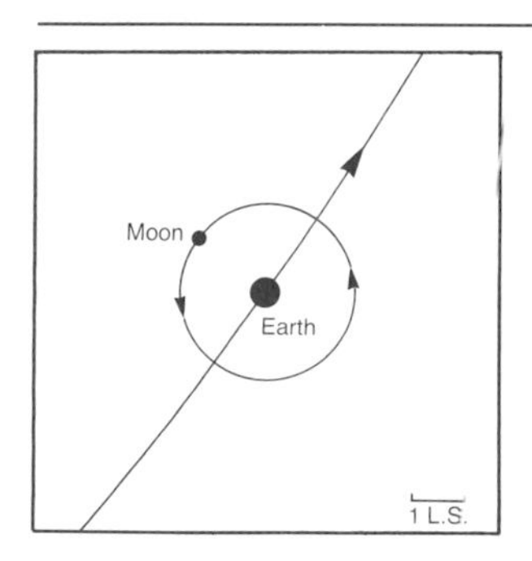

The small diagram at left is the first in a series of 10 that shows our cosmic voyage outward from Earth. Here, we see the moon's orbit around Earth. A small section of the Earth's orbit around the sun is also shown. L.S. = light-second.

Earth and its moon, facing page, as they might appear to an approaching spaceship. The moon's rugged, cratered face still retains the scars of its youth. Because the moon has no air or water on its surface—and probably never did—there is nothing to erode the big craters.

The craters on the moon and on other worlds in the solar system were blasted out by huge chunks of debris left over after the planets formed, as seen in the artist's rendering above.

STEP TWO: 4 LIGHT-MINUTES FROM EARTH

The nearest worlds beyond the moon are Venus and Mars. Their distances from Earth vary as they orbit the sun, but they are always several light-minutes away. Astronauts may visit Venus someday, but it won't be soon. Venus's thick atmosphere has made its surface hotter than a barbecue grill. Mars is a much friendlier place.

Sometime in the 21st century, humans will stand on Mars and gaze back toward Earth, the brightest "star" in the Martian sky. Even from the nearest planets, Earth is a mere dot in the heavens – just as the other planets appear to us.

Explorers on Mars will probably wear spacesuits like those used on the moon by the Apollo astronauts. They will communicate with Earth by radio. However, two-way conversations will be complicated. Radio signals from Earth, which travel at the same speed as light, take several minutes to go from Mars to Earth. Therefore, Mission Control will hear the astronauts several minutes after they speak. It will then be several more minutes before the astronauts receive a reply from Earth.

The problem will be even more serious for future explorers of Jupiter's moons. They will have at least a one-hour wait between sending a message to Earth and hearing an answer. From Pluto, the most distant planet, one-way communication will take four hours.

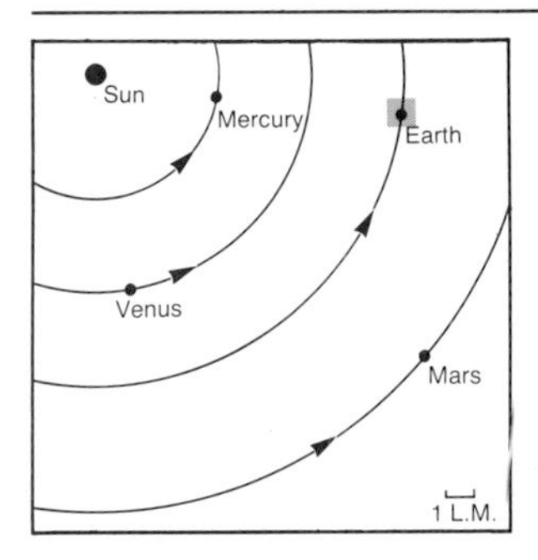

The diagram at left shows the orbits of the four inner planets in our solar system. The shaded box is the sector covered by the Step One diagram on page 7. L.M. = light-minute.

The scene on facing page depicts the view back toward Earth from the surface of Mars shortly after a Martian sunset. Earth is the bright, white "star." The moon, nestled beside Earth, is much fainter. A telescope would be needed to see oceans or land on Earth through breaks in the clouds. Phobos, the larger of Mars's two satellites, is the potato-shaped body.

The real Mars, seen in the photograph above, taken by the American Viking spacecraft in 1976, is a windswept desert world with sand dunes and a dusty yellow sky.

STEP THREE: 4 LIGHT-HOURS FROM EARTH

Our voyage outward from Earth takes us past the giants of the solar system – Jupiter, Saturn, Uranus and Neptune – to the realm of Pluto, the last of the known planets. At this distance, four light-hours, Earth is an insignificant starlike dot beside the remote sun. It is so cold on Pluto that the planet's atmosphere is frozen on the ground. Ice is everywhere. Although the sun appears as small as a pinhead in the sky, it still sheds enough light so that future explorers will have no trouble seeing Pluto's landscape. Future astronauts will be able to walk on Pluto if they wear properly insulated spacesuits. However, a voyage from Earth to Pluto would take at least 10 years using the best rockets now available. The trip will probably not be made until faster spaceships are designed.

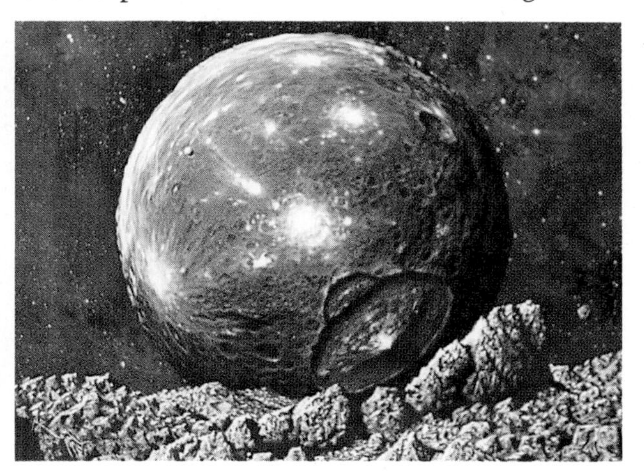

Pluto has the largest orbit of any planet – it takes 248 Earth years to complete one trip around the sun – but it is not always the most distant planet. A section of its orbit is closer to the sun than Neptune's orbit. From 1979 to 1999, Pluto is eighth and Neptune ninth from the sun. They will never collide, however.The two orbits are like a highway interchange. One goes over, the other under.

Recent measurements have shown that Pluto is by far the smallest planet – smaller even than the Earth's moon. In 1978, astronomers were surprised to find a moon almost as big as Pluto itself orbiting that planet. The moon has been named Charon. Pluto is the only planet in the solar system that will not be examined close up by a space probe during the 20th century.

Searches for planets beyond Pluto have been unsuccessful. If such worlds do exist, they are either much smaller than Pluto or extremely far away and thus difficult to detect. Pluto may be the outermost planet, but there are other members of the sun's family deeper in the abyss – the comets.

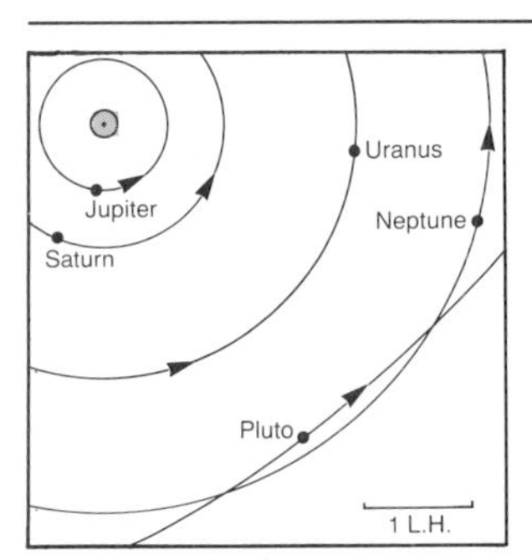

Orbits of the outer planets are shown in the diagram at left. A section of Pluto's orbit is closer to the sun than Neptune's orbit is. The shaded box is Step Two. L.H. = light-hour.

The icy landscape of Pluto, facing page, is illuminated by the distant sun. Pluto is the smallest planet, no bigger than the Earth's moon, but has the largest orbit about the sun.

No spacecraft has ever been to Pluto, but the illustration above shows what the planet might look like from the surface of its satellite, Charon. Charon is more than half the size of Pluto. The bright splashes on Pluto were caused by meteorite impacts long ago. Pluto is composed mainly of ice, darkened somewhat by dust and dirt.

STEP FOUR: 2 LIGHT-MONTHS FROM EARTH

Beyond Pluto is the realm of the comets. Comets are city-sized chunks of frozen material – mostly water ice – orbiting in the frigid cold far from the sun. Astronomers estimate there are at least a trillion comets roaming the void several light-months from the sun. Even at that distance, the sun's gravity is strong enough to keep most of the comets from floating off to the stars.

Comets orbit the sun like miniature planets. But the orbits are enormous. One trip around the sun could take a million years. Unlike the planets, which follow nearly circular paths, comets orbit in sausage-shaped ellipses. One end of the ellipse can be as close to the sun as Pluto or Neptune, while the other can extend up to a light-year out into space.

Most comets remain much as they were billions of years ago. Occasionally, however, a comet passes close to one of the four giant planets, which causes its orbit to shift and sends it nearer to the sun. If a comet comes within Mars's orbit, it begins to melt due to the solar heating. Gases emerge and form a cloud. Solar radiation pushes the cloud back into a streaming tail. Viewed from Earth, the curving, filmy tail of a bright comet can be spectacular.

Although there are a tremendous number of comets, there is plenty of room for them. On average, they are several light-minutes apart. That is about the same as the distance between our planet and Mars or Venus. The outer solar system is by no means crowded.

Other stars likely have huge families of comets too. Trillions of comets may have escaped from their stars to roam the vast interstellar spaces.

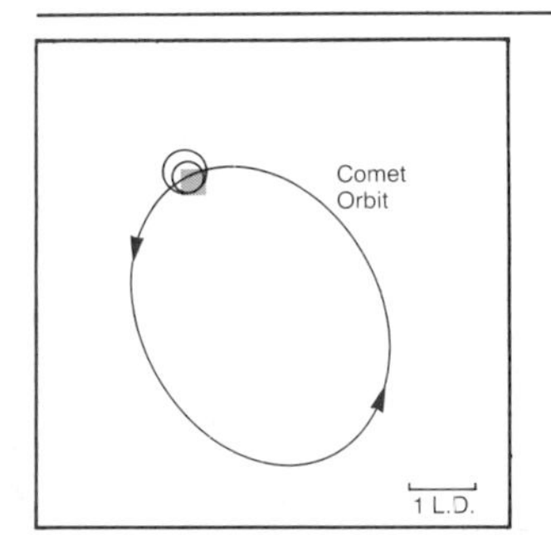

Comet orbits are usually oval-shaped and are much larger than the orbits of the planets. One comet orbit is shown in the diagram at left. Many comet orbits are bigger than this. Shaded box is Step Three. L.D. = light-day.

Comets are huge islands of ice drifting in gigantic orbits that carry them up to a light-year from the sun. Most of them, like the one illustrated on facing page, remain farther from the sun than Pluto. Out here, in the frigid depths of space, no comet tail can exist. The comet's spongy, prickly appearance stems from the way ice crystals combine in space.

When a comet does venture close to the sun, the heat melts the surface ice and gases burst out, as shown in the artist's rendering above.

STEP FIVE: 4.3 LIGHT-YEARS FROM EARTH

Even a beam of light, which hurtles from Earth to the moon in 1.3 seconds, takes more than four years to reach the nearest star, Alpha Centauri. New types of rockets will have to be developed to span the enormous distances to the stars. They will use completely different fuels from those that power the space shuttle and other launch vehicles today. Some scientists say that such flights will never be made; others, that starflight is inevitable. Whatever the future holds, it is highly unlikely any space missions will reach Alpha Centauri during our lifetimes.

Telescopes on Earth can tell us a great deal about the stars. We know, for instance, that Alpha Centauri is actually three stars: one almost identical to our sun, one a bit dimmer and one very faint. The faint one, Proxima Centauri, circles the other two in a huge million-year orbit. Proxima happens to be on the side of its orbit that places it slightly closer to us than the other pair (4.24 light-years compared with 4.34 light-years). Planets might exist around each of the three stars in the Alpha Centauri system. The sky from a planet circling the brightest of the three is pictured on facing page.

When astronomers refer to Alpha Centauri, they mean all three stars in the system, which can be seen through a telescope. Many other stars that appear single to the unaided eye have companions.

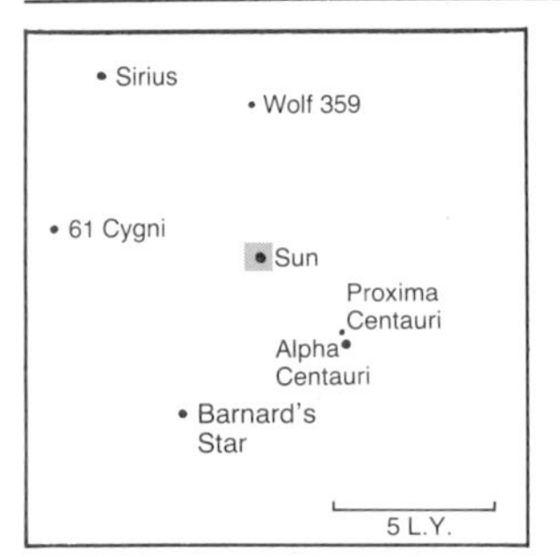

Diagram at left shows some of the stars closest to the sun. Shaded box is Step Four.

The view on facing page is an imaginary scene of a planet in the Alpha Centauri star system. The planet is in the foreground. Beyond it is a moon and a second sun—Alpha Centauri is actually a three-star, or three-sun, system. The primary sun is below the horizon and to the right. The Earth's sun is the bright star in the Milky Way Galaxy.

An artist's rendering, above, shows the second nearest star, a dim red sun called Barnard's Star, as seen from one of the moons of a planet similar to Jupiter. At one time, astronomers thought they had discovered such a planet orbiting Barnard's Star, but it appears to have been a mistake.

STEP SIX: 10,000 LIGHT-YEARS FROM EARTH

Our sun is a citizen of a vast wheel-shaped star city called the Milky Way Galaxy. All of the stars seen on a dark night are part of the same galaxy. The sun is situated about 25,000 light-years from the galactic core. The galaxy is fat at its centre, or nucleus, but thin elsewhere. Astronomers call the central region the nuclear bulge. It appears bright because stars are closer together there than in the sun's region. The nuclear bulge is 10,000 light-years from top to bottom. Out in the "suburbs," where the sun resides, the galaxy is 3,000 light-years thick.

All of the stars in the galaxy are moving. Most move together, like a huge group of joggers running around a giant racetrack. That's how the galaxy keeps its flattened shape. The sun takes about 200 million years to complete one trip around the nucleus. Its path is almost circular. Many of the stars in the sun's vicinity have similar orbits. The sun has gone less than halfway around the galaxy since the time when dinosaurs ruled Earth.

In Step Five, we saw how far apart the stars really are. However, once we move thousands of light-years away and look back, most of the galaxy's stars are not seen as individuals but blend together into giant curved arms of starlight. The same effect occurs during a snowfall of big, fluffy flakes. We see the individual flakes floating close by, but in the distance, they all blend into a haze.

Some of the galaxy's stars are outside the main wheel-shaped section. They are called halo stars. Beyond them are two small galaxies known as the Large Magellanic Cloud and the Small Magellanic Cloud, named after the Portuguese explorer Ferdinand Magellan who observed them during one of his sea voyages. At the time, nobody knew what they were, but since they look like clouds, that is what they were called. The two Magellanic Clouds are satellites of our galaxy. Step Seven takes us out to the large cloud's vicinity.

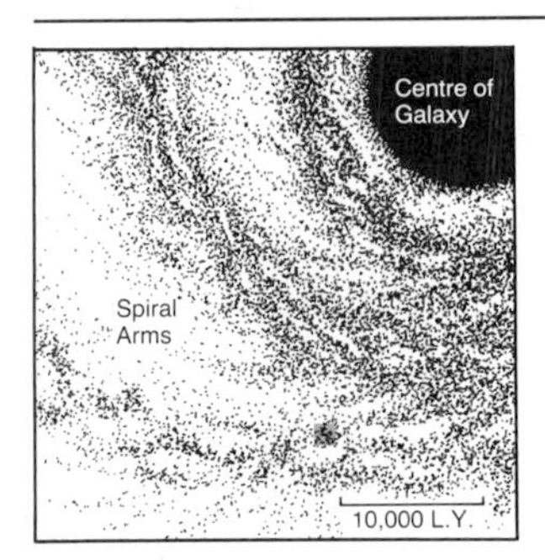

The sun is located about two-thirds of the way out from the nucleus of the Milky Way Galaxy on the inside edge of one of the spiral arms. Shaded area in diagram at left is Step Five.

Our sun is one of 200 billion stars belonging to the Milky Way Galaxy. The scene on facing page depicts the sector of the galaxy that includes the sun. Place your thumb on the picture, near the bottom. You are now covering all of the stars that are seen in the sky on a dark night—only a tiny portion of the entire galaxy.

The photograph above shows another galaxy like our own. The Milky Way Galaxy would look like this if viewed from deep space. The galaxy is seen edge-on. For a face-on view of a similar galaxy, see page 23.

STEP SEVEN: 170,000 LIGHT-YEARS FROM EARTH

Suppose the distance to Alpha Centauri, the nearest star system, were just two city blocks – a short walk. How far would it be from one edge of the Milky Way Galaxy to the other? Forget about walking. A one-week car ride all the way across North America would get us only a little past the halfway point.

Our galaxy is gigantic. Here is another example of its vastness: a large sports stadium seats 50,000 people. If all the stars in the galaxy were divided equally among them, each person would have several million stars. Now, a million is *lots* of stars. Just saying the word "star" a million times would take about a week (without stopping for food breaks or even sleep). If starships are ever developed, there will be no shortage of destinations.

We do not know exactly what our galaxy looks like from the outside because we are on the inside. We see it all around us as the Milky Way, that misty band arching overhead most evenings during the year. However, astronomers have determined that the galaxy is wheel-shaped and is about 80,000 light-years from edge to edge. Thousands of galaxies are close enough to us that large telescopes have revealed their structures. Some are spirals like our galaxy, others are football-shaped and are called elliptical galaxies. A third type, the irregular galaxies, have no particular shape.

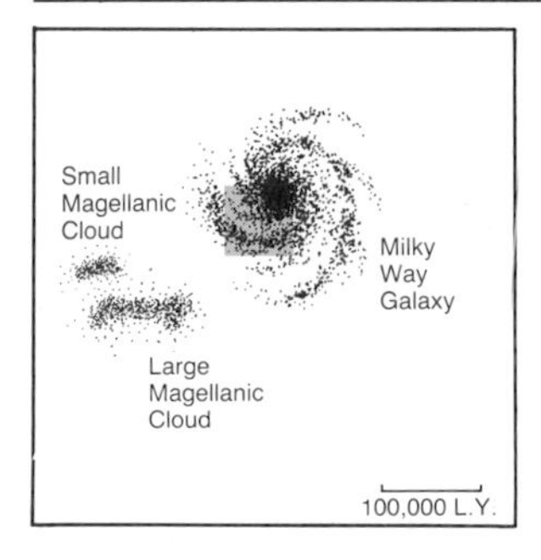

Our galaxy is a huge spiral galaxy. The Large Magellanic Cloud, the nearest galaxy to us, is a satellite of the Milky Way Galaxy. Shaded box in the diagram at left is Step Six.

Earth, sun and planets were all born about 4½ billion years ago in a nebula like the one in the photograph above. Nebulas are clouds of gas in the spiral arms of galaxies. Several nebulas can be seen as small pink patches along the arms of the galaxy in the photograph on page 23.

The dramatic scene on facing page is a panoramic view of the Milky Way Galaxy as seen from an imaginary planet of a star in the Large Magellanic Cloud. The Large Magellanic Cloud is a small galaxy, the nearest one to the Milky Way.

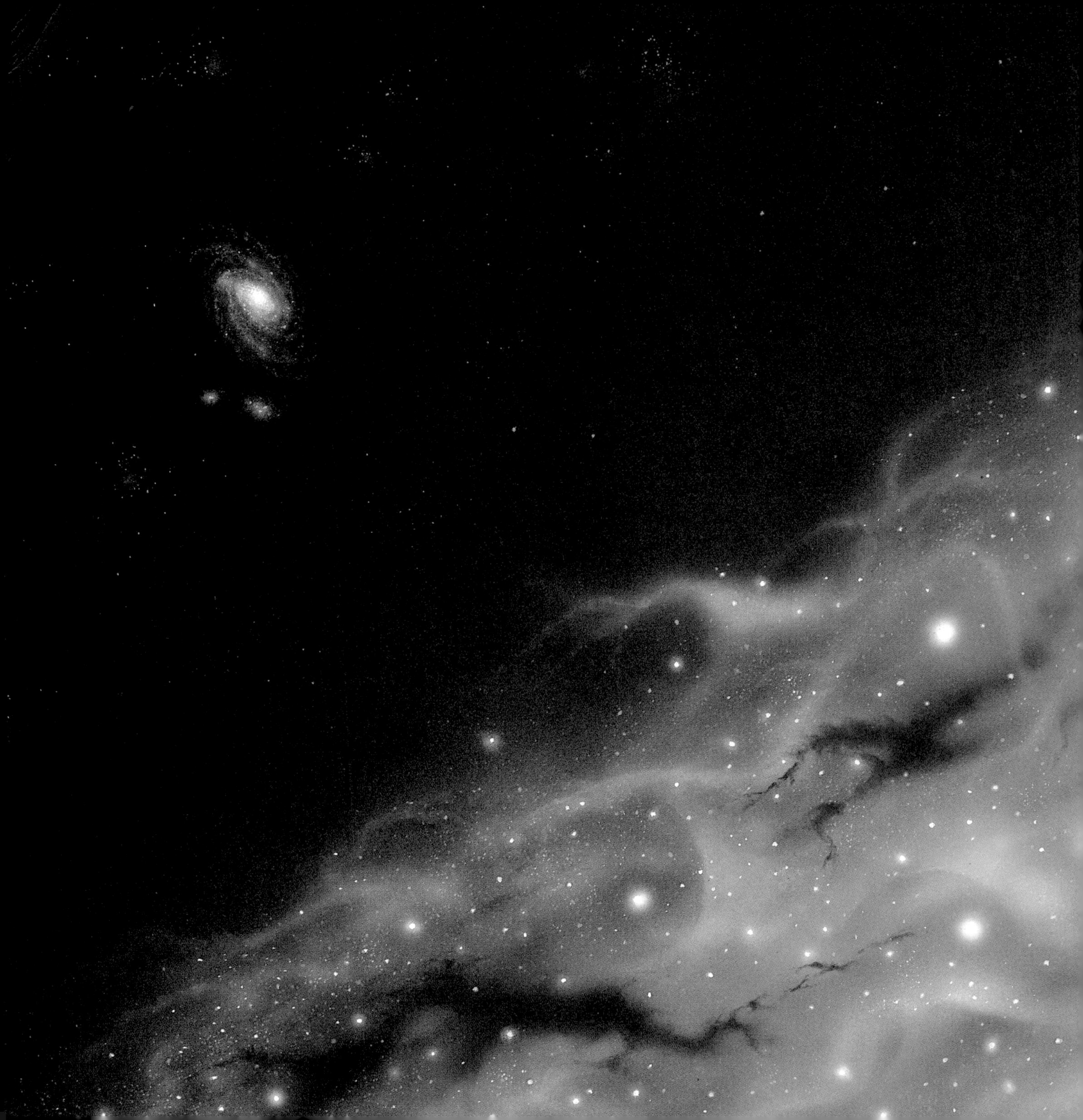

STEP EIGHT: 2,500,000 LIGHT-YEARS FROM EARTH

The spaces between stars seem immense, but they are microscopic compared with the distances between galaxies. Imagine that the distance from the sun to Alpha Centauri is only as far as this book is from your eyes as you read these words. On such a scale, the Andromeda Galaxy, the nearest major galaxy similar to the Milky Way, would be 150 miles (240 km) away.

The Andromeda Galaxy is 2.3 million light-years from our galaxy. It is a spiral galaxy slightly larger than the Milky Way. A bit farther away, at 2.5 million light-years, is a smaller spiral called the Triangulum Galaxy. The view on facing page shows what our galaxy might look like from the vicinity of the Triangulum Galaxy. The Milky Way, Andromeda and Triangulum galaxies are the largest of about 30 galaxies that form what astronomers call the Local Group – our galactic neighbourhood.

Galaxies shaped like the Milky Way Galaxy are known as spiral galaxies. Their elegant arms are composed of billions of stars. Our sun is located on the edge of a spiral arm. The spiral shape suggests that galaxies turn – and they do. Stars near the nucleus make one revolution in 50 million years. The sun goes around once in 200 million years, which means it has made less than one revolution since the first dinosaurs roamed the Earth's surface.

The arms of spiral galaxies do not seem to twirl up or unwind over long spans of time. Instead, they maintain their shape by continuously re-forming. How? Astronomers are still trying to unravel the detailed secrets of the beautiful spiral shapes.

Between galaxies, there is almost pure emptiness. Atoms of hydrogen gas float here and there, but a whole roomful of intergalactic space would contain only about a dozen atoms. Long ago, perhaps 15 billion years or more, most of the gas in the universe gathered into clumps that became galaxies. Then the gas in the galaxies collected to form stars. Stars are still being born in galaxies today.

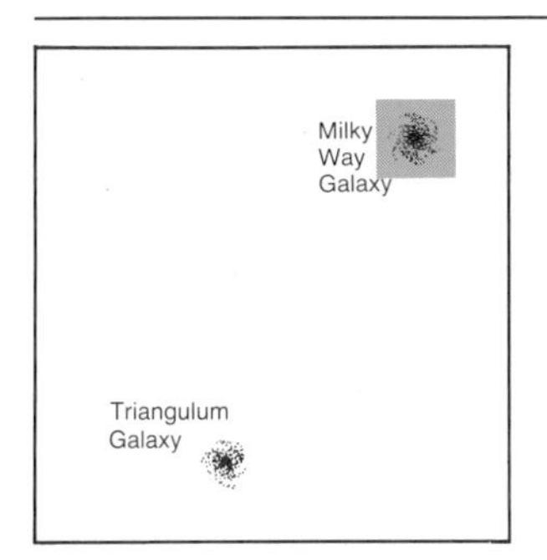

Our galaxy is the second largest galaxy in the Local Group. Shaded area is Step Seven.

The Whirlpool Galaxy, seen in the artificially coloured photograph above, is similar to both our galaxy and Andromeda, but its arms are spread farther apart. The bright patch at the end of one arm is a companion galaxy.

The Milky Way Galaxy and its two companions, the Magellanic Clouds, float in the abyss as seen from the outskirts of the Triangulum Galaxy in the illustration on facing page. The space between galaxies is almost pure emptiness. A few nearby dwarf galaxies are seen in the blackness along with some more distant galaxy groups. The loneliest places in the universe are the vast gulfs between galaxies.

STEP NINE: 10,000,000 LIGHT-YEARS FROM EARTH

Galaxies are islands of stars in the great void of space. The Milky Way and the other galaxies that form the Local Group are all within three million light-years of each other. Beyond them is a gap of several million light-years before other groups of galaxies are encountered.

Individual galaxies vary greatly in size. Small ones have less than a million stars – about the same number as the population of a fairly large city. The smallest known galaxy is the Carina Dwarf, a neighbour of the Milky Way. It contains just a few hundred thousand stars. The biggest galaxies have trillions of stars.

Distances to even the nearest galaxies are enormous. The light we see from the Andromeda Galaxy, for example, has been on its way to Earth for more than two million years.

The five largest galaxies in the Local Group are, in order of size, Andromeda, Milky Way, Triangulum, Large Magellanic Cloud and M110 (a companion to Andromeda). The Andromeda Galaxy contains at least 300 billion stars, the Milky Way Galaxy 200 billion. These are by far the heavyweights. Triangulum is next, with about 15 billion stars, followed by the Large Magellanic Cloud and M110, with about 10 billion each.

In describing distances and sizes in the universe, very big numbers must be used. A billion is almost unimaginably large. A billion seconds is 31 years. A billion one-dollar bills would completely fill the average school gymnasium. A billion raindrops would overflow an olympic-sized swimming pool. And a trillion is 1,000 times a billion!

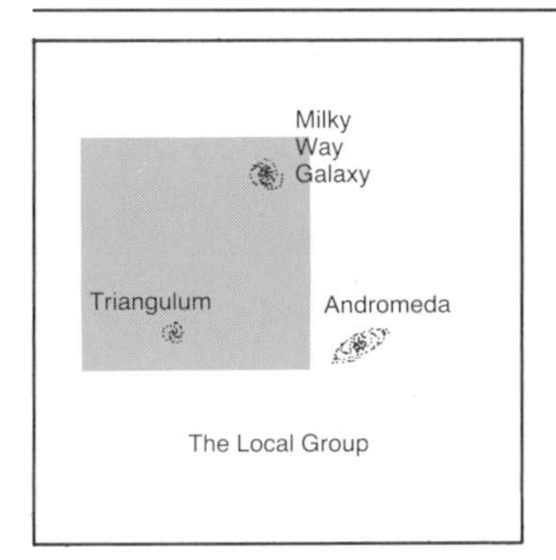

The Andromeda Galaxy is the nearest large spiral galaxy to the Milky Way. Shaded area in diagram is Step Eight.

The photograph above shows a spiral galaxy millions of light-years from our galaxy. The spiral arms are blue because the brightest stars in them are hot blue suns. Most of the stars in the central region are yellow. The pink patches in the arms are nebulas.

The three largest galaxies of the Local Group—the Milky Way, Andromeda and Triangulum—are seen on facing page as they would appear from a distance of about 10 million light-years. All three are spiral galaxies. The Andromeda Galaxy is the largest galaxy in the Local Group. It is 2.3 million light-years from the Milky Way.

STEP TEN: 300,000,000 LIGHT-YEARS FROM EARTH

Our voyage outward will end here, 300 million light-years from Earth. We could go farther, but the scenery would be basically the same for billions of light-years. The Milky Way Galaxy is so remote that it is just a dot in the background. In every direction, galaxies are scattered across the blackness. However, they are not spread randomly throughout space like motes of dust floating in the air. Instead, galaxies tend to clump into clusters like the Local Group. These clusters, in turn, are collected into chains and ribbons called galaxy superclusters, which are separated from one another by enormous voids of empty space.

Examination of the galaxy clusters has shown that the universe is expanding. Each cluster is moving away from its neighbours. For example, the Centaurus galaxy cluster is 200 million light-years from us and is hurtling away so fast that every two seconds, it increases its distance from us by the width of North America. We do not know whether the universe will continue to expand forever or whether the outrush will eventually stop.

Because the universe is expanding, it cannot have existed forever. A billion years ago, the galaxy clusters were closer together. A billion years before that, they were closer together still, and so on. By working backwards in this way, astronomers have calculated that the universe began about 15 billion years ago. They call the beginning the Big Bang – the creation of the universe.

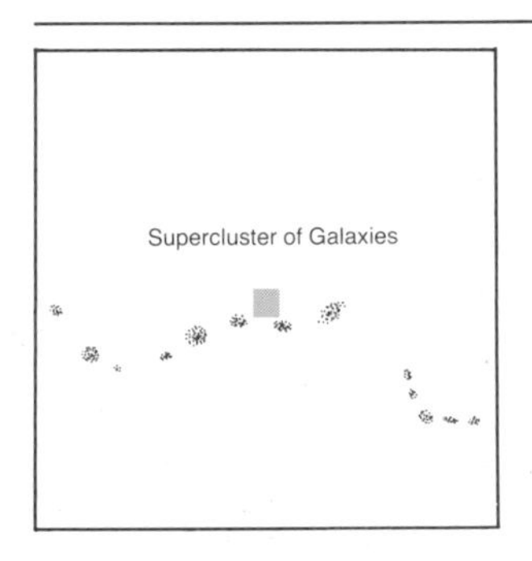

In diagram at left, the area covered by Step Nine is shown as a small shaded box.

The universe of galaxies is so vast that the Milky Way Galaxy is just an insignificant blip in this view on facing page from 300 million light-years away. The millions of galaxies are not spread evenly in space but are collected into clusters, chains and ribbons. In this scene, thousands of distant galaxies are mere dots.

Some galaxies are not spirals and have no graceful curving arms of stars. The galaxy in the photograph above may be the remains of two galaxies that collided more than a billion years ago. Astronomers suspect that many galaxies without spiral arms were once involved in galaxy collisions.

THE SOLAR SYSTEM: OUR SUN'S FAMILY

The nine planets, their five dozen moons, millions of asteroids and trillions of comets are all members of the solar system. The sun is at the centre, its gravity ruling the orbital paths of its children.

A model of the solar system gives an idea of its size and the sizes of its various members. Let's use a major-league baseball stadium located in the centre of a large city for the model. The sun, the size of a baseball, rests on home plate. Mercury, Venus, Earth and Mars, each about the dimensions of the ball in a ball-point pen, are, respectively, 1/8, 1/5, 1/3 and 1/2 of the way to the pitcher's mound. A pea near second base is Jupiter. In shallow centre field is a smaller pea, Saturn. Uranus, the size of this letter O, is at the fence in deep centre field. Neptune and Pluto, a letter O and a grain of salt in our model, are just outside the park.

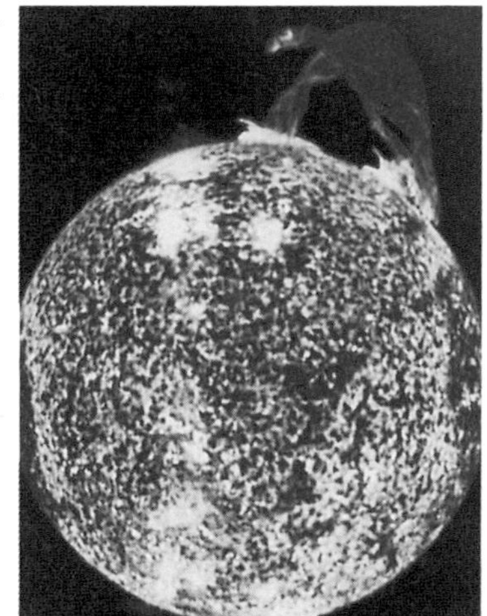

There is more to the solar system. Between Mars and Jupiter, just inside the bases, are millions of bits of sand and dust – the asteroids. Ranging beyond Pluto, right to the edge of the city, are trillions of comets. All of them would be microscopic in size. The nearest star would be a baseball in another city more than 1,000 miles (1,600 km) away.

Dozens of space probes, with names like Voyager, Venera, Mariner and Zond, have given us close-up looks at the planets and their moons. They are amazing and beautiful worlds. They divide into two distinct groups: big and small. The four giant planets – Jupiter, Saturn, Uranus and Neptune – are many times larger than the other five planets. Neptune alone contains far more matter than the combined mass of Earth, Venus, Mars, Mercury, Pluto and all of their satellites.

The four giant planets are different in another way. They are mainly gas, rather than rock. For this reason, they are called gas giants, or Jovian planets, after Jupiter, the largest of the group. But even the gas giants pale in comparison with the ruler of them all – the sun, our nearest star.

The sun is so huge that more than a million Earths could fit inside it. One hundred Earths could be lined up side by side across its middle. The temperature at the sun's surface is 11,000 degrees F (6,000 °C) – hotter than a torch that can melt steel. The surface is a churning sea of gas that receives its energy from the sun's interior. That energy is dispersed into space as light and other forms of radiation.

The sun's "furnace" is deep inside at the core, where hydrogen is turned into helium by enormous pressures and temperatures of 55 million degrees F (30 million °C). The same reaction generates the power of a hydrogen bomb. The energy produced by the reaction at the core takes millions of years to work its way up through the bulk of the sun to the surface. All of the stars seen in the night sky are basically like the sun. Some are hotter, some cooler. A few stars are more than 100 times larger than the sun, while others are no bigger than Earth.

THE SUN AND ITS PLANETS

	Diameter (Earth = 1)	Length of Year	Known Moons
SUN	109	–	–
MERCURY	0.38	88.0 days	0
VENUS	0.95	224.7 days	0
EARTH	1.00	365.3 days	1
MARS	0.53	687 days	2
JUPITER	11.2	11.9 years	16
SATURN	9.4	29.5 years	17
URANUS	4.0	84.0 years	15
NEPTUNE	3.8	164.8 years	8
PLUTO	0.2	247.7 years	1

	Distance From Sun (Earth's distance = 1)	Length of Day (sunrise to sunrise)
SUN	–	–
MERCURY	0.39	176 days
VENUS	0.72	117 days
EARTH	1.00	24 hours
MARS	1.52	24h 39m
JUPITER	5.2	9h 50m
SATURN	9.5	10h 39m
URANUS	19.2	17h 18m
NEPTUNE	30.1	16h 03m
PLUTO	30 to 49	6d 9.3h

The sun, facing page, source of light and energy for Earth, is our local star. Giant prominences leaping from the surface are sometimes taller than the distance from Earth to the moon. They are caused by intense, twisting magnetic fields at the solar surface.

A family portrait of the Earth's relatives. The sun is so huge that only a segment fits into the illustration at right. The diagram would have to be as large as 10 football fields to show distances and sizes accurately.

VENUS AND MERCURY: TWO BROILED WORLDS

The two planets that are closer to the sun than Earth is have very different surface conditions, but they have one thing in common: they are both *hot*. Mercury is an airless world like the moon. Because it is three times nearer the sun than our moon is, its surface bakes at temperatures far above the boiling point of water. Astronauts could, however, still explore the planet using well-insulated, air-conditioned spacesuits. They would see a landscape of craters, ridges and rubble-covered plains much like the lunar surface that the Apollo astronauts walked on.

Venus orbits the sun about midway between Mercury and Earth. It is the closest planet to Earth and almost identical to our planet in size. For many years, astronomers thought conditions on Venus might be similar to those on Earth, although a bit warmer because it is closer to the sun. In the 1960s, space probes flew by Venus and discovered the real story: Venus's surface is hotter than Mercury's. It is hot enough to melt lead. A kitchen oven at its maximum setting reaches about 550 degrees F (290°C). That would be a deep-freeze on Venus, where a thermometer would read 860 degrees F (460°C) both day and night. Why is it not cooler at night? Venus's thick atmosphere of carbon dioxide acts like a blanket to keep the heat in. The atmosphere absorbs heat from the sun and releases it back into space very slowly, so the temperature everywhere on Venus is the same.

Some astronomers say Venus was not always this way. It once had oceans, but the hotter sunshine (because Venus is closer to the sun) caused so much water to evaporate that the atmosphere became like the roof and walls of a

greenhouse, which trap heat. When the greenhouse effect got started, the oceans boiled and Venus was left with a thick, cloudy atmosphere.

In 1975, two Russian space probes landed on Venus. They survived long enough in the tremendous heat to transmit photographs of the surface back to Earth. The pictures show a rocky plain. Sunlight is dimmed by the clouds as on a heavily overcast day on Earth. There are canyons, volcanoes, ridges and plains, as well as a few craters. But it will be a long, long time before humans ever explore them.

The photograph at lower left shows Venus as it would appear from an approaching spaceship. Thick haze at the top of the dense atmosphere blocks the view of the planet's surface.

The planet Mercury bakes under the glare of the sun, which shines 10 times as intensely as it does on Earth, in the illustration above. Mercury's day is twice as long as its year. A scorching-hot day lasts 88 Earth days, followed by a night of equal length. The temperature range during that cycle is almost 1,000 degrees F (600°C).

Humans will not be taking a stroll on Venus in the foreseeable future. In the artist's rendering, top left, a robot probe is descending through the atmosphere, which is as thick as water and hundreds of degrees hotter than boiling oil. An explorer would have to wear a diving suit far more sophisticated in design than any now in existence. The surface rock on Venus is hot enough to melt lead. Poisonous acid vapours are mixed in with the carbon-dioxide atmosphere.

MARS: THE MOST EARTHLIKE PLANET

Mars has always fascinated astronomers because it, of all the planets, seems to be the most like Earth. About 100 years ago, many astronomers were convinced that Mars was inhabited. Their telescopes revealed dark areas that varied in intensity with the seasons, similar to the way forests and other vegetation change from one season to the next on Earth. They saw straight, dark lines linking the dark regions. These "canals," it was suggested, were built by Martians to transport water over the cool, dry planet.

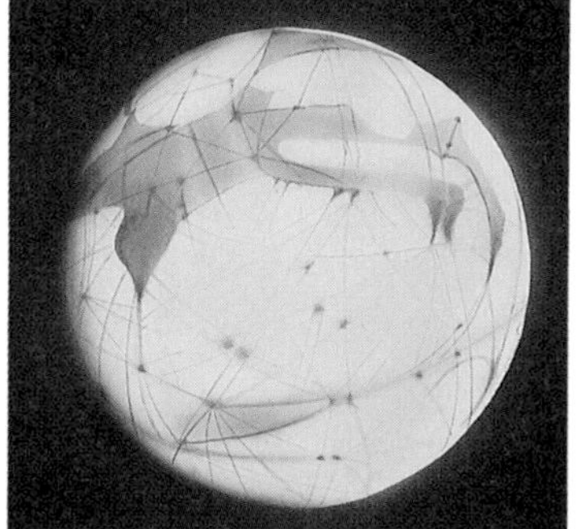

We now know that the canals were the result of eyestrain on Earth, not intelligence on Mars. The shifting dark areas are caused by dust storms, rather than vegetation. Mars is a vast desert. Its surface dirt is reddish orange because it contains rust. Winds of up to 250 miles per hour (400 km/h) whip dust particles into the atmosphere, making the sky a permanent peach colour.

The Martian atmosphere is too thin for humans to breathe. In any case, it is mainly carbon dioxide, the gas humans exhale. At night, the temperature drops to minus 90 degrees F (-70°C) at the equator and even lower elsewhere. Daytime temperatures are seldom above freezing. Yet some features of Mars resemble those on Earth.

Mars has dozens of gigantic volcanoes, many of them far larger than any on Earth. The biggest one, Olympus Mons, can be seen on the right side of Mars in the illustration on facing page. Olympus Mons is as big as the island of Newfoundland and three times higher than Mount Everest. A few clouds can be seen on the left side of Olympus Mons. Clouds are rare because Mars has so little water vapour in its atmosphere. Some areas of Mars are heavily cratered and resemble the lunar landscape. Elsewhere, there are hundreds of channels that are almost certainly ancient riverbeds. Water once flowed on Mars, but the planet's surface dried up long ago.

If there ever was liquid water on Mars, then there may have been life. However, two spacecraft that landed on Mars in 1976 found no evidence of life as we know it.

Mars has craters like the moon and polar ice caps like Earth. More Earthlike features include volcanoes, dust storms and ancient dried riverbeds. At one time, long ago, water flowed on Mars. Today, the planet is a frigid desert, though water may be frozen underground. The illustration at right shows Mars as it might appear from near the surface of Phobos, one of two small moons that orbit Mars.

The canals of Mars are shown in the drawing above, made 80 years ago. The straight-line features were believed to be waterways built by intelligent Martians, but they proved to be optical illusions. Even so, the idea that some form of life might be discovered on Mars still persists. The prospects are not good, however, because two Viking landers failed to find a trace of life in samples of Martian soil. Future plans for the exploration of Mars include a wheeled vehicle that will travel across the Martian deserts for months. Eventually, humans will follow.

JUPITER: KING OF THE PLANETS

Jupiter makes the rest of the planets in the solar system look puny. The king of the planets is more massive (that is, it contains more material) than all of the other planets combined. If it were possible to drive around Jupiter's equator in a car, the journey would take six months of nonstop 24-hour-a-day travel. A similar nonstop drive around the Earth's equator would take only two weeks.

Yes, Jupiter is gigantic. And it is also alien. The planet is mostly hydrogen gas; there is no place to land. A spaceship descending into the atmosphere would sink deeper and deeper until the pressure of the thickening gases crushed it like a paper cup. Deep inside Jupiter, the atmosphere is so dense it resembles a hot liquid. Deeper still is a rock and metal core, perhaps the size of Earth.

The top of the atmosphere is a colourful blanket of swirling clouds. The white clouds are ammonia ice crystals. The beige and brown clouds are ammonium hydrosulphide ice crystals. The highest clouds in the Earth's sky are ice crystals too – water ice.

Jupiter is the fastest-spinning of the planets. Its day is less than 10 hours long. The rapid rotation stirs the atmosphere and clouds into dark- and light-coloured stripes, called belts and zones. The turbulence in the clouds often develops into giant storms, some the size of North America. The biggest storm, known as the Great Red Spot, is larger than the entire Earth.

The planet's family of 16 moons is like a miniature solar system. Four of the moons – Io, Europa, Ganymede and Callisto – are giants. Ganymede is the largest moon in the solar system and is bigger than the planet Mercury. The other three are almost as large. The four moons are so big they can be seen in binoculars – like little stars nestled beside the brilliant image of Jupiter. Each night, they change position as they orbit the big planet.

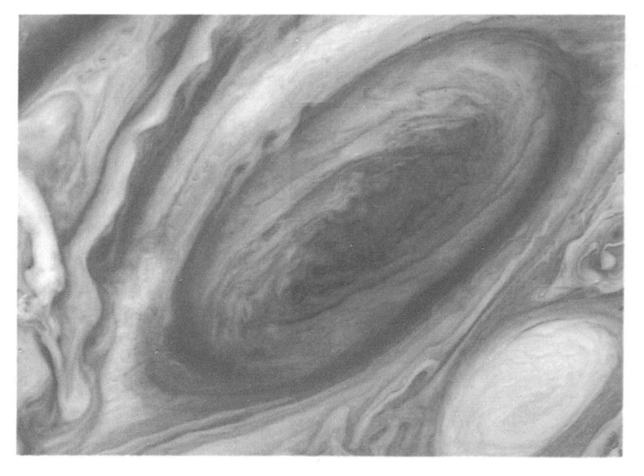

Future exploration of Jupiter by robot spacecraft will pave the way for human exploration of the giant planet, perhaps toward the end of the 21st century. Whenever it happens, the first expeditions will land on the moons instead of plunging into Jupiter's ocean of clouds. That deadly voyage will likely be left to robots.

If explorers ever venture to the top of Jupiter's ocean of clouds, the view might look like the illustration at right. Winds here reach speeds of 270 miles per hour (430 km/h). Below the clouds is an atmosphere of hydrogen and helium.

The Great Red Spot in Jupiter's atmosphere, seen in the photograph above, is a giant storm that is larger than the entire Earth. The picture was taken by the American Voyager 2 space probe as it sped past Jupiter in July 1979.

SATURN AND BEYOND: RINGS AND ICE WORLDS

Few sights in the universe can match the beauty of Saturn and its spectacular system of rings. The rings are composed of trillions of snowballs, some the size of a house, others no bigger than specks of dust. Saturn is 10 times farther from the sun than Earth is, so the temperature there is always far below freezing. Ice in the rings never melts. According to one theory, the rings formed when two of Saturn's moons collided. The pieces continued to smash into each other, creating smaller and smaller chunks until there were so many that they now appear as a solid mass.

Each particle in Saturn's rings has its own orbit around the planet, like a tiny moon. There are more particles zooming around Saturn in some parts of the ring structure than in others, which gives the rings a grooved appearance like that of a record. The rings are the thinnest structure known in nature when width is compared with thickness. A model of the rings as thick as one page in this book would be larger than a football field. Saturn's rings could almost span the distance from Earth to the moon, yet they are only as thick as the height of a 30-storey apartment building.

Saturn has 17 known moons, more than any other planet. The largest, Titan, is the size of the planet Mercury. Its atmosphere is about the same density as the Earth's and consists of nitrogen (95 percent) and methane (5 percent). Titan's surface is solid ice, possibly covered with an ocean of liquid methane and ethane – substances known as natural gas on Earth. Future explorers of Titan may need boats in which to get around. They will also need lights. Titan's atmosphere is heavily smog-laden. Daylight is barely brighter than full moonlight on Earth.

Beyond Saturn are two more giant planets, Uranus and Neptune. Both are smaller than Saturn but are still much bigger than Earth. Uranus has a system of thin rings of refrigerator-sized ice chunks. Because they are so skinny, Uranus's rings are difficult to detect and were not discovered until 1977. Neptune's rings were found in 1989.

The many divisions in Saturn's rings are seen in the photograph at left, taken as the Voyager 1 spacecraft hurtled past the planet in 1980. Part of Saturn is seen at upper left.

Sometime in the 21st century, humans will explore the moons of Saturn. In the illustration at right, the ice cliffs of Saturn's moon Mimas are scaled by an expedition from Earth. In the background looms Saturn, with its rings casting an inky shadow on the planet's ammonia-ice-crystal clouds.

PLANETS OF OTHER STARS

Four hundred years ago, the philosopher Giordano Bruno said: "Innumerable Earths circle around other suns." The idea was not very popular at that time. Bruno said a few other unpopular things, and he was burned at the stake in February 1600. Today, scientists think Bruno was probably right. But we still do not know for certain whether planets of other stars exist. No object like Jupiter, Neptune, Earth or any of the rest of the planets in our solar system has been seen orbiting another star. The planets are probably there, but they are just too dim to be detected.

A planet shines by reflecting the light of its sun. That is what makes the moon and the planets in our solar system appear bright. But the moon and planets are nearby. Stars are many thousands of times more distant. When astronomers search for planets near other stars, the star's light overwhelms the feeble reflected light of any planets that might happen to be there. It is like trying to see a tiny fly hovering near a streetlight. From far away, the insect is invisible even though the light is easily seen. This problem has prevented a sighting of even a big planet like Jupiter.

However, astronomers are confident that they are getting close to finding planets of other stars.

One researcher has taken photographs which indicate that several red dwarf stars are wobbling slightly in their paths across the sky. The wobbling may be caused by the gravity of planets bigger than Jupiter. But this is what scientists call indirect evidence. It cannot be called a discovery until one of these suspected planets is actually seen.

Looking elsewhere in space, astronomers see clouds of gas and dust collapsing to form stars. Computers have been used to predict the way this would happen. These simulations, as they are called, suggest that planet formation is a natural process occurring at the same time a star is born. Other astronomers have found discs of dust and debris surrounding some stars – something like Saturn's rings only much larger. The discs could be leftover material from the formation of planets or material just beginning to form into planets.

More powerful telescopes are being built every year. The first positive discovery of a planet of another star should be made soon. Today, we can only imagine what other solar systems look like.

The illustration at left shows a hypothetical gas giant planet three times as massive as Jupiter. It rules a family of dozens of moons, including one the size of Earth (foreground). A system like this could exist, but no planets beyond our solar system have yet been confirmed.

The red dwarf star Proxima Centauri, accompanied by a planet and the planet's moon, are illustrated above. Such renderings are based on what could be, not what is known for sure.

NEARBY STARS: OUR SUN'S NEIGHBOURS

There are approximately 1,000 stars within 55 light-years of the sun. About 950 of them are smaller and much dimmer than the sun. That means the sun is not really an average star, since it is larger and brighter than 95 percent of the stars in our neighbourhood. Faint stars called red dwarfs are average stars. The dimmest known red dwarf star, LHS 2924, shines so feebly that if it were to replace our sun, daylight on Earth would be no brighter than a night with a full moon.

The nearest star visible from Canada and the United States is Sirius. It is also the most brilliant star in the night sky. Sirius is 8.6 light-years from Earth. It is 23 times as bright as the sun and twice as big. If Sirius were in the sun's place, the Earth's surface would be hotter than Mercury's is today.

Orbiting Sirius at the same distance that Uranus orbits the sun is a tiny star about the size of Earth. It shines white-hot. Such a star is known as a white dwarf. Some stars collapse when they have used up most of their atomic fuel. White dwarfs are a type of collapsed star. Sirius's companion has the same mass as our sun, which means its matter is very densely packed. A teaspoonful of white-dwarf material would weigh as much as a car.

When astronomers began studying the nearby stars in detail, they discovered that many of them are not alone. Stars orbit other stars just as the moon orbits Earth. The Alpha Centauri system has a dimmer sun that orbits a sunlike star once every

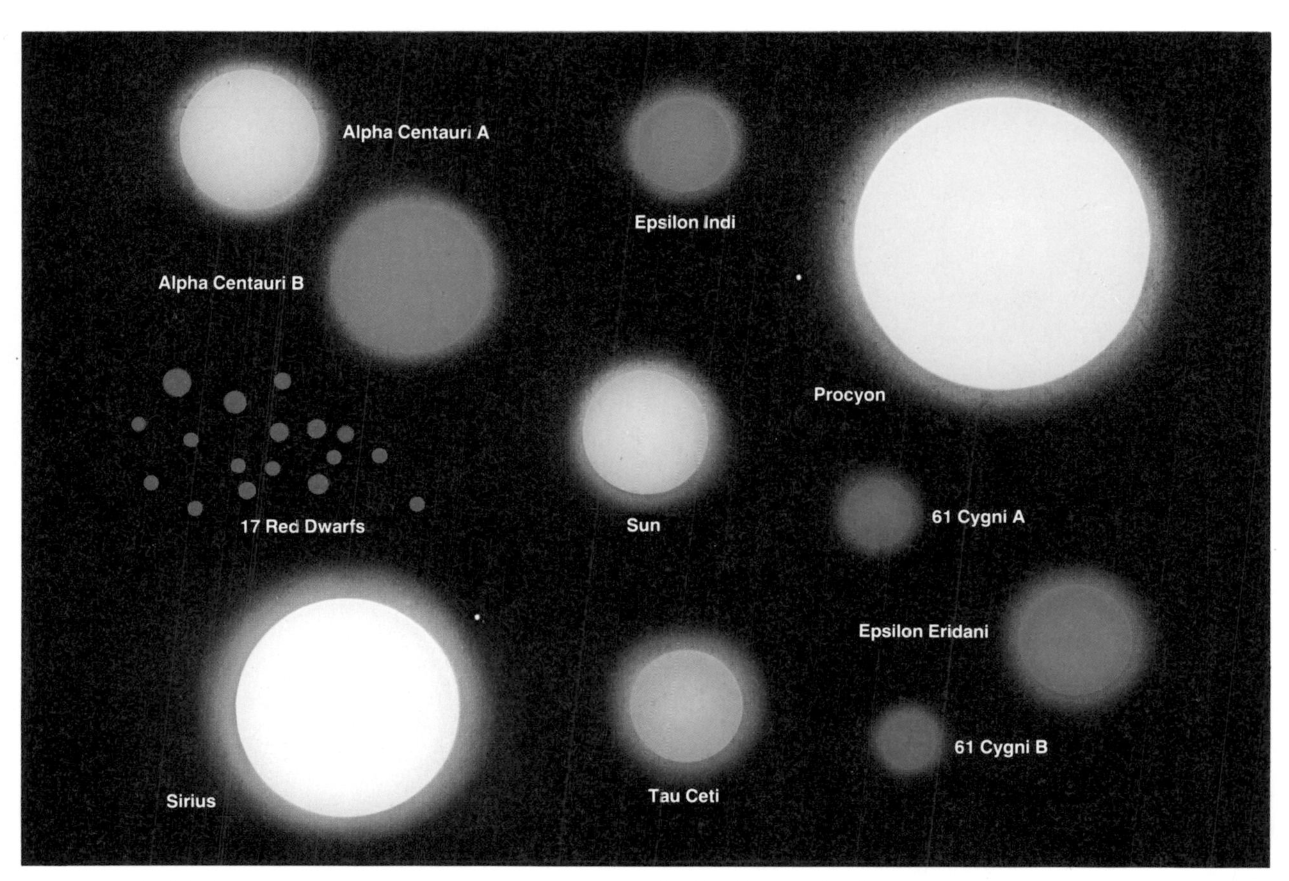

80 years. A third star, Proxima Centauri, orbits the other two. Almost half of the stars visible in our night sky have other stars as companions. Maybe more than half do, but we will need better telescopes to see them.

All the known stars within 12 light-years of the sun are shown grouped together in the illustration above. Red dwarfs are by far the most common type of star. Sun-like stars are less common, and stars brighter than the sun are fairly rare. Procyon and Sirius are the only nearby stars significantly brighter and larger than the sun. The small stars beside them are white-dwarf companion stars.

The view from a planet with two suns might look like the artist's rendering on facing page.

DISTANCES TO THE NEAREST STARS

Star	Distance from sun in light-years
Proxima Centauri	4.24
Alpha Centauri A & B	4.34
Barnard's Star	5.97
Wolf 359	7.7
BD + 36°2147	8.2
L726-8 A & B	8.4
Sirius A & B	8.6
Ross 154	9.4
Ross 248	10.4
Epsilon Eridani	10.8
61 Cygni A & B	11.1
Epsilon Indi	11.2
BD + 43°44 A & B	11.2
Procyon A & B	11.4
Tau Ceti	11.8

HOW STARS END THEIR LIVES

The sun has been shining for roughly five billion years. It has not changed much during that time. It may have gradually brightened, but it looks about the same today as it did when it shone on the dinosaurs. Big changes are in store for the sun in the future, however.

Five billion years from now, the thermonuclear furnace at the sun's core will have used up much of its hydrogen fuel. During billions of years of energy production, the sun has been converting hydrogen to helium. Instead of going dark, the sun will have a final fling of radiance when it "burns" helium. The sun's core will then become hotter, and its outer layers will expand like a balloon blowing up. Instead of its present width – 100 times the Earth's diameter – the sun will grow to 10,000 times our planet's diameter and become a red giant. The sun will be red because its surface will be cooler than it is now. But there will be so much surface, and therefore so much more heat, that Earth will be cooked to a cinder.

After several hundred million years, the outer portion of the red giant star will be expelled to form a bubble around the core. The core, meanwhile, will shrink to a white dwarf star such as Sirius's companion.

Stars more massive than the sun end their lives with a bang. After they become red giants, a sudden explosion, called a supernova, blasts the star apart; expanding clouds of matter are sometimes all that remain. Or the star's core could collapse into a neutron star or a black hole. A neutron star is the size of a mountain, with two or three times as much material packed into it as there is in our entire sun. Neutron stars are far denser than white dwarf stars. A teaspoonful of neutron-star matter, if brought to Earth, would weigh more than a million railroad locomotives.

The artist's rendering at left shows the scorched planet of a red giant star. The sun will look like this five billion years from now. Red giant stars exist for a few hundred million years at most – not long compared with a star's total life span.

The giant ring of gas in the photograph at right is the last gasp of a star like the sun. After the red-giant stage, material is puffed off in an expanding bubble that is illuminated by the hot stellar core seen at centre. The star then shrinks to a white dwarf.

BLACK HOLES: GRAVITY WHIRLPOOLS IN SPACE

Black holes are probably the weirdest objects in space. They are created during a supernova explosion. If the collapsing core of the exploding star is large enough – more than four times the mass of our sun – it does not stop compressing when it gets as small as a neutron star. The matter crushes itself out of existence. All that remains is the gravity field – a black hole. The object is gone. Anything that comes close to it is swallowed up. Even a beam of light cannot escape.

Like vacuum cleaners in space, black holes suck up everything around them. But their reach is short. A black hole would have to be closer than one light-year to have even a small effect on the orbits of the planets in our solar system. A catastrophe such as the swallowing of Earth or the sun is strictly science fiction. However, someday far in the future, when we can travel to a black hole, its space-warping gravity may prove useful to boost the speed of a spaceship.

The nearest known black hole, Cygnus X-1, is 10,000 light-years from Earth. It was discovered because it orbits around a large blue star and is swallowing gases emitted by the star. As the gases descend into the hole, they are swirled into a whirlpool by the spinning gravitational field. It is like a giant bathtub drain in space. This heats the gas before it disappears into the hole. The energy given off by the heating is in the form of x-rays, which can be detected by Earth-orbiting satellites. A few other black holes have been found in the same way. There must be many more black holes that do not have material swirling into them and therefore are invisible. They could be detected only by their gravitational effects on nearby objects.

The illustration above shows a black hole near a giant blue star. The black hole itself cannot be seen. It is surrounded by material that will soon plunge into the hole. A black hole not consuming anything would be completely invisible.

Matter swirling into the gravitational whirlpool of a black hole known as Cygnus X-1 is seen in the illustration at right. As the hole orbits a giant star, it sucks in gases. In a few million years, the black hole may "eat" most of the blue star.

QUASARS: THE BEACONS OF DEEP SPACE

Quasars are the brightest objects in the universe. Some shine more brilliantly than the combined light of 1,000 galaxies like our Milky Way. Because they are so bright, they can be seen at tremendous distances. Quasars are the farthest objects ever observed; some are so remote that their light began its journey here long before Earth even existed.

For many years after the discovery of the first quasar, in 1963, astronomers tried to determine what makes quasars so bright. Now, they think they have the answer. A quasar is thought to be the result of a collision of galaxies.

A galaxy collision is not like an automobile crash, where bits and pieces go flying in all directions as the cars are brought to a sudden halt. Because the stars that make up galaxies are very far apart (as we noticed in the first section of this book), entire galaxies can pass through one another. None of the stars in one galaxy would ever come in contact with any stars in the other galaxy. It's something like the crossover manoeuvre during a precision drill with foot soldiers or mounted police. However, the galaxies become twisted out of shape. Just as the moon's gravity causes tides on Earth, one galaxy's gravity pulls on the other. Eventually, the galaxies can have their spiral arms torn away and their shapes severely distorted, like the pair of colliding galaxies shown in the photograph at lower right. Sometimes, the two galaxies never part, and their stars merge into a single giant galaxy such as the one pictured on page 25.

How do colliding galaxies create a quasar? All large galaxies probably have very massive black holes at their centres. When galaxies collide, the huge nebulas of gas and dust within one galaxy pass right over the massive black hole in the other galaxy. This abundant supply of "food" begins to swirl into the gravity whirlpool surrounding the black hole. Matter whirling around the hole is heated to fantastic temperatures. Huge amounts of radiation then flood into space. A quasar is born.

Galaxies do not crash into each other very often today because the universe is expanding. As it expands, the space between galaxies increases, so there is now less chance of galaxies bumping into each other than there was in the past. Billions of years ago, when the galaxies were closer together, collisions were more common, and therefore, quasars were more common.

Most quasars are billions of light-years away. It

The photograph at lower right shows the collision of two galaxies. Each galaxy has been severely twisted out of shape by the gravitational pull of the other galaxy. These galaxies will never return to their former appearance. The universe holds many examples of wrecked galaxies.

A quasar is caused by a massive black hole at the centre of a galaxy. If one could be seen close up, it might look like the illustration at upper right. Gases from a galaxy collision have swirled into a flat disc about one light-month across. At the centre of the disc is the black hole, one billion times the mass of the sun. The hole acts like the drain of a sink as matter plunges into it. So much material is available that some overflows and is ejected in a jet like the spray from a fire hose. The tremendous heat from all this activity is melting the planet in the foreground. Several stars near the quasar are being destroyed as well (they look like comets). Quasars are the most powerful sources of radiation in the universe.

has taken billions of years for their light to reach us. Because quasars are so powerful, they can be seen at greater distances than normal galaxies. Quasars are the most remote objects in the known universe. Some of them are thought to be 10 to 12 billion light-years from Earth.

Distance is like a time machine. We see the universe not as it is but as it was. The light from a star 50 light-years away is seen on Earth 50 years after it left the star. So we see the star as it was 50 years ago. In the same way, light from quasars shows us what they were like billions of years ago.

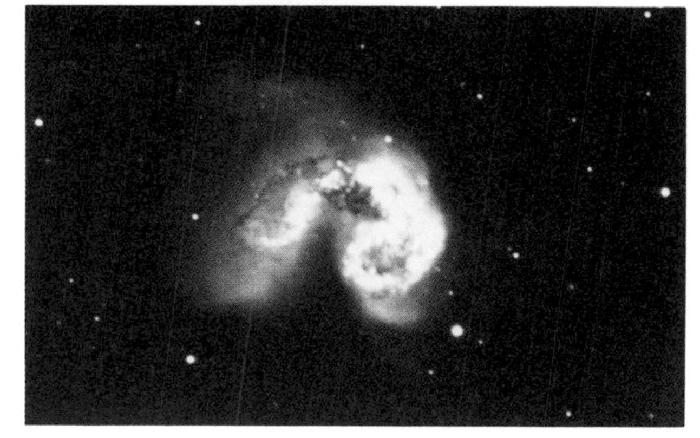

EXTRATERRESTRIALS: IS ANYONE OUT THERE?

Is Earth the only place in the universe with intelligent life? Most people believe we are not alone. "There must be other creatures," they say. "After all, there are so many stars and the universe is so vast." Perhaps you have said something like this. But just because most people believe something does not mean it is true. Scientists need proof, and so far, there is not one scrap of solid evidence to support the existence of extraterrestrials.

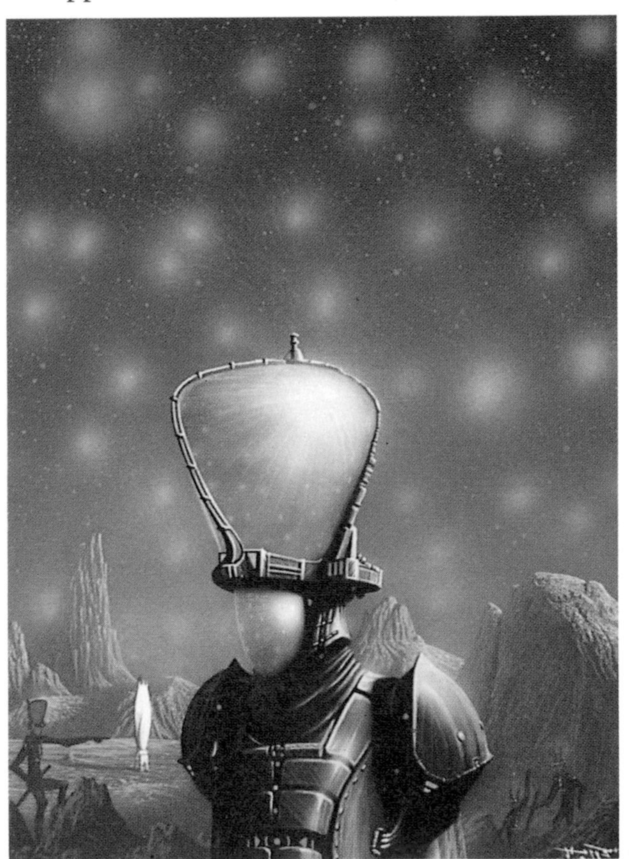

Proof that extraterrestrials exist would be one of the greatest discoveries of all time. There is so much that we would want to know: What do they look like? Where is their home planet? Do they have wars and disease and suffering?

In 1984, a radio telescope in Massachusetts began 24-hour-a-day scanning of the heavens, searching for a signal from some other civilization. So far, nothing has turned up. Scientists say our radio telescopes could detect a message beamed in our direction from the other side of the galaxy. They plan to continue the search for at least the rest of the century. But would aliens be sending messages to Earth? We simply do not know.

Whenever extraterrestrials are mentioned, the subject of UFOs usually comes up. Thousands of UFO (unidentified flying object) sightings are on record. Virtually everyone knows someone who has seen one. UFO sightings at night almost always turn out to be aircraft, satellites, meteors, searchlights or bright stars and planets near the horizon. They are "unidentified" only because the people seeing them do not know what they are. This is understandable because few people other than astronomers and pilots spend much time looking at the night sky. A bright object that seems to hover near the horizon flashing all the colours of the rainbow might surprise someone glancing out the kitchen window, but an astronomer knows it is probably the planet Venus or the star Sirius because such behaviour is displayed almost every night.

There are *some* UFO sightings that defy explanation. But we must be careful not to assume that extraterrestrials are involved just because some-

thing unusual has been observed. This is not proof. Clear photographs of an alien device would be excellent evidence, but no such picture has ever been published.

Most of us would like to believe that we are not alone in this vast universe. However, we are likely alone in one sense. Even if there are many other intelligent civilizations on planets of other stars, they won't be like us. Not only will their appearance be different (which doesn't really matter), but they will be at a different level of development (which *does* matter).

The universe is billions of years older than Earth. There has been plenty of time for the development of civilizations much older than us. Some of these alien civilizations will be as far in advance of us as we are beyond the level of the cave people that were our ancestors. Even if they are only a few hundred years beyond us in development, the difference between our technology and theirs will be unimaginable. The more advanced they are, the more likely it is that they already know about us. We would look very primitive to them. For this reason, they may have decided not to contact us.

In the illustrations above, two planets of other stars are shown. One is slightly less massive than Earth, the other more massive. The smaller planet has about half the surface gravity of Earth. In this low-gravity environment, many creatures have evolved into flying animals. On the larger planet, the surface gravity is 1½ times the Earth's. Here, most life has evolved in the oceans instead of on land. Astronomers do not know if such worlds exist. However, there is no reason why they could not exist. We simply do not know if there is any form of life beyond Earth.

Are we alone in the universe? Are there other intelligent creatures exploring other worlds? Such questions cannot be answered at present. It may be hundreds or thousands of years before we meet our cosmic cousins from other planets. If we are alone, the search will continue forever because we'll never know for sure. However, it is fun to speculate about the appearance of aliens as in the illustration on facing page.

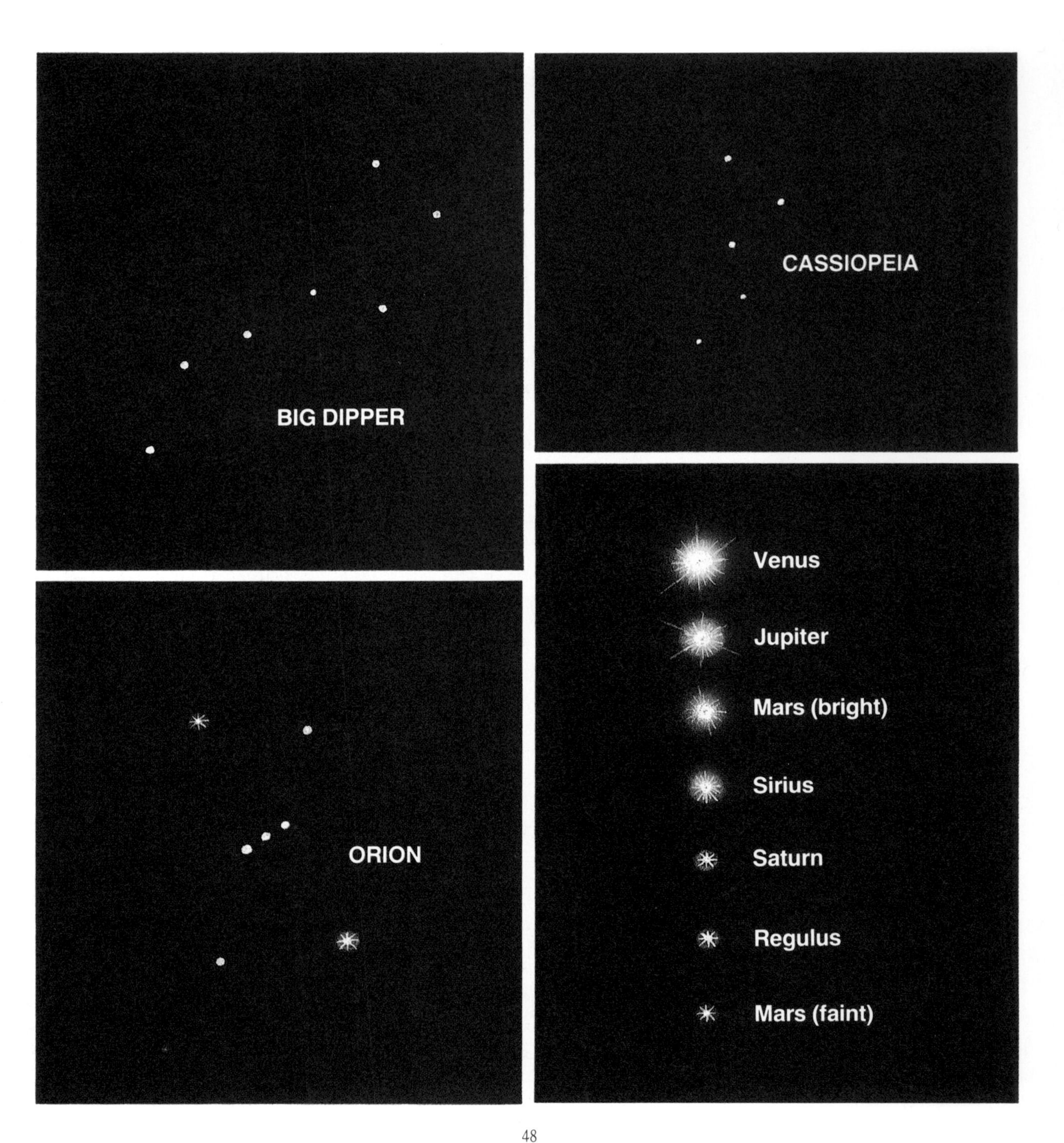
BIG DIPPER
CASSIOPEIA
ORION
Venus
Jupiter
Mars (bright)
Sirius
Saturn
Regulus
Mars (faint)

HOW TO RECOGNIZE PLANETS, STARS AND CONSTELLATIONS

Stargazing is easy. It can be done anywhere. All it requires is our eyes and a dark night sky. In the city, street lamps and other types of lighting flood the sky, making it impossible to see faint stars, but most of the brighter stars are visible. The charts in this section show only bright stars and are useful from any location in Canada or the United States on any clear night of the year.

The hardest part is recognizing a star or constellation for the first time. Constellations are patterns of stars, often named after mythological characters or animals. Some constellations look like what they were named for, but most do not. Leo, for example, appears a bit like a lion, but Pegasus – no matter how you examine it – does not resemble a flying horse.

To make that first sighting as easy as possible, let's start with the most familiar star pattern: the Big Dipper. You may know it already – its saucepan shape is unmistakable. The Big Dipper is not a true constellation but, rather, a modern name for the seven brightest stars of Ursa Major, the great bear.

The Big Dipper is always visible in the northern sky from Canada and the northern half of the United States. In the southern states, it is too close to the horizon to be seen easily in the autumn months but can be observed during the rest of the year. The diagram on the next page shows how the Big Dipper changes position in the sky as the seasons pass, due to the Earth's motion in its yearly orbit around the sun.

Cassiopeia, which looks like the letter W, is seen in the same section of the sky as the Big Dipper and is almost as easy to recognize. In autumn, when the Dipper is low, Cassiopeia is near overhead and is the chief signpost to autumn stars. The beautiful constellation Orion, with its "belt" of three bright stars in a row, is the winter sky's landmark.

Planets seldom twinkle, whereas stars almost always glitter like diamonds. Although planets are *not* larger than stars, they appear larger in our night sky because they are so much closer to us – just as someone standing next to you looks taller than a tree far off in the distance. (Your eye alone cannot detect the difference in size between planets and stars, but binoculars or telescopes will reveal it.) Therefore, planet light is less likely to be interrupted by the wavelike turbulence that is always present in the air. It is turbulence that causes twinkling.

The easiest way to recognize your first star is to identify three prominent star patterns: the Big Dipper and the constellations Orion and Cassiopeia. Each has a distinctive shape and acts as a pointer to other star groups. The Big Dipper is the most useful. Orion always appears as shown, but the Dipper and Cassiopeia can be upside down or angled compared with their orientations here.

Some of the planets are brighter in our night sky than Sirius, the brightest star. Venus, a dazzling white, is the brightest of all. It can be seen in the western sky after sunset during several months each year. Jupiter is brighter than the stars too. Mars is a distinct rusty orange colour, its brightness changing from month to month as its distance from Earth varies. Because planets are always moving in their orbits around the sun, their positions cannot be plotted on our sky charts. However, they travel only along a specific track called the ecliptic. This "path of the planets" is marked in pale red on the charts.

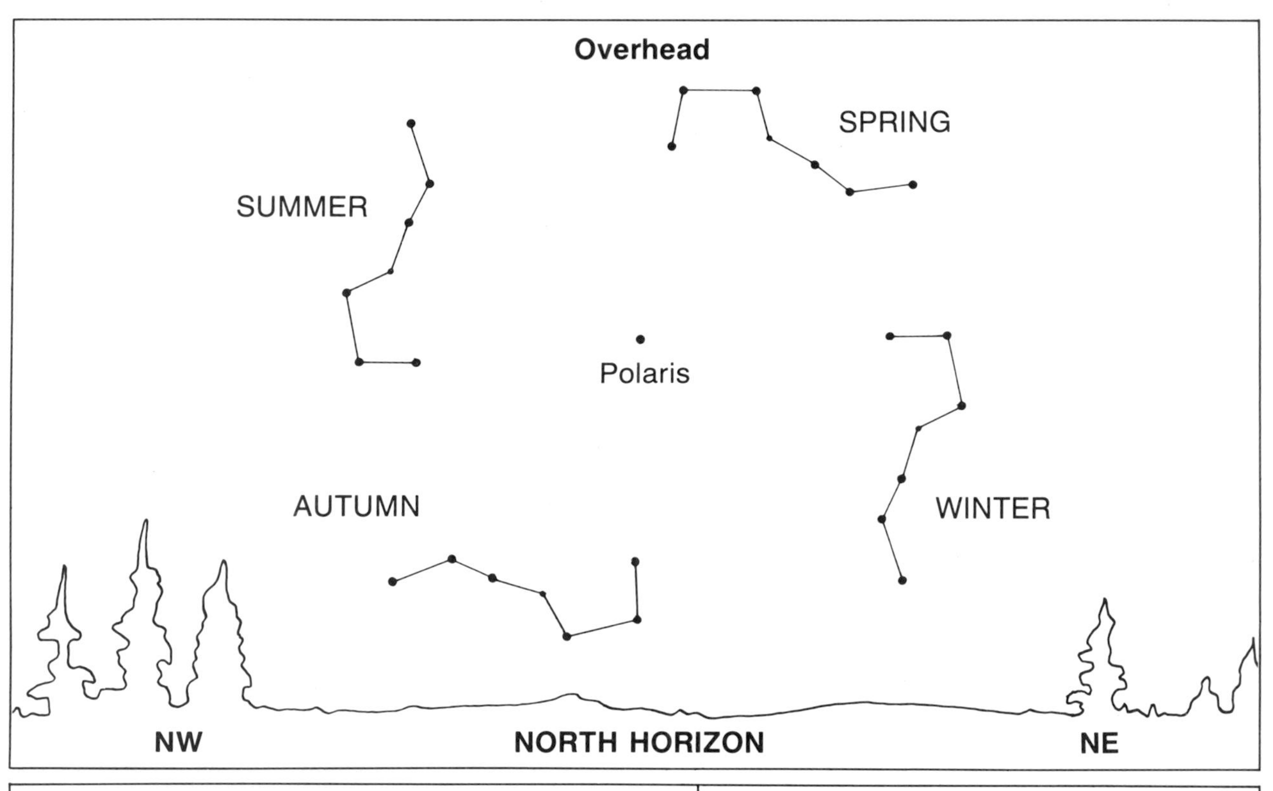
Overhead
SPRING
SUMMER
Polaris
AUTUMN
WINTER
NW
NORTH HORIZON
NE

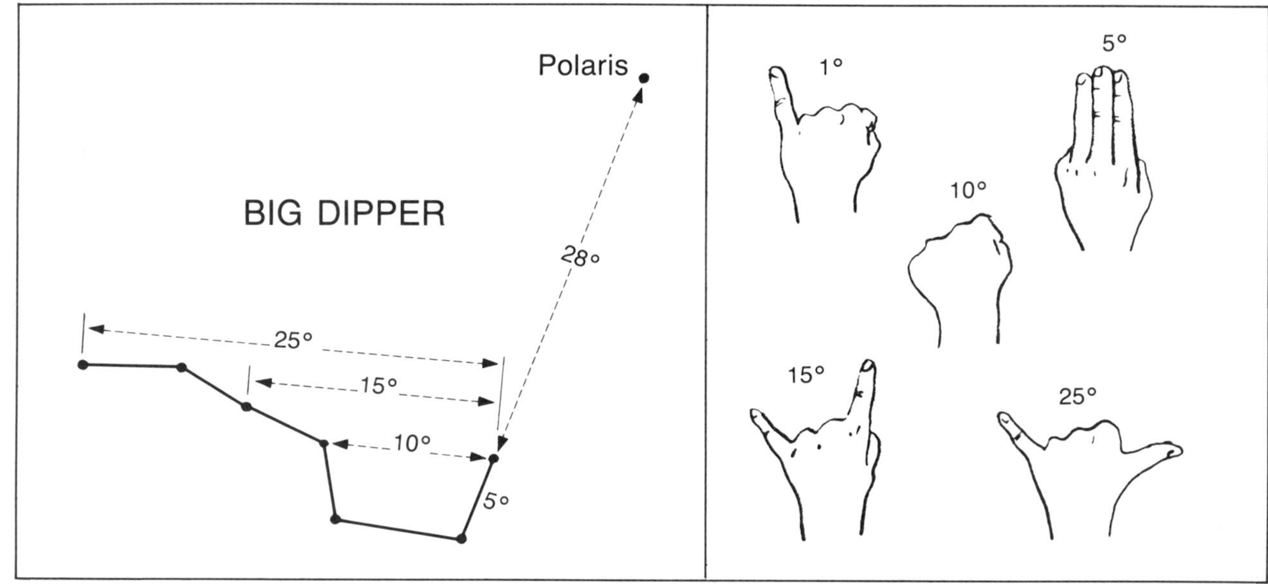
Polaris
BIG DIPPER
28°
25°
15°
10°
5°
1°
5°
10°
15°
25°

EASY SKY GUIDES: LEND A HELPING HAND

Before you go outside, it will help to know how big the Dipper appears in the sky as well as the sizes of some of the other star patterns we will be observing. To find this out, we need some sort of measuring stick. Distances between stars in the night sky are given in degrees. One degree is about the width of the end of your little finger when your arm is held out straight. Three fingers are 5 degrees, and a fist is 10 degrees. If you spread your fingers as wide as possible, the Big Dipper is roughly as long as the distance from the tip of your thumb to the end of your little finger. The Dipper's bowl is as wide as a fist, or 10 degrees.

The Big Dipper is 25 degrees long. Ninety degrees is the distance from the horizon to straight overhead. By referring to the diagram on facing page and using your fingers and hand, you can estimate any distance from 1 degree to 25 degrees. These measurements have nothing to do with real distances. The idea is to use hand measurements as an aid to finding the constellations. The hand-measuring system works for both children and adults because smaller hands are attached to shorter arms, so the proportions remain the same.

The illustrations on the following pages depict the night sky as realistically as possible. If observing from a dark country location, you will see more stars than are shown. From the city, fewer will be seen. But that doesn't matter, because we will be talking about the brightest stars.

Each illustration shows a section of the sky 90 degrees high and 90 degrees wide. The overhead portion of the sky is at the top. The illustrations are accompanied by smaller finder charts showing exactly the same stars, with lines connecting the main star patterns and with the names of the brightest stars. Locater arrows guide you from the key groups – the Big Dipper, Orion and Cassiopeia – to the other stars.

You are now ready to begin stargazing. On the next cloudless night, check the Big Dipper diagram on facing page for the position of the Dipper, then go outside and identify it. Next, find the main chart for the current season (there may be more than one). The finder chart notes the times when it works best, although it is usually useful an hour before or after these times. Read the summary under the guide chart for observing tips. If you are having problems, it may be because you are not beginning with the Big Dipper or because the Big Dipper is not on that particular chart. You must then find Orion or Cassiopeia to get started. Look at their shapes on the charts, and try again. Once you have the first pattern, the rest is easier, and you are on your way to knowing the stars.

One possible obstacle remains. A planet lurking somewhere along the ecliptic (the path of the planets) may cause confusion. Jupiter, Mars and Saturn are the most likely culprits. Times when they are in the chart areas are given in the text. Planets will *never* be seen more than two or three degrees from the ecliptic.

The position of the Big Dipper shifts during the year—from close to the northern horizon in autumn to nearly overhead in spring. The change is gradual and is due to the Earth's yearly orbit around the sun. Polaris, the North Star, is the only star that seems to stand still throughout the year. Using your hand "ruler," you can measure sky distances that help gauge the sizes of constellations and the distances between them. The Big Dipper, for example, is as wide as a hand fully outstretched.

THE BIG DIPPER AND THE SUMMER SKY

Throughout the entire summer, the Big Dipper hangs in the evening sky in the northwest, ready to act as a guide. The Dipper is the stargazer's most useful stellar pattern. Its seven stars serve as fingers to point the way to stars and constellations all year long. In summer months, Polaris, Cassiopeia, Vega and Deneb can be found by using the convenient locater arrows pictured at right.

The Big Dipper is not entirely a chance alignment of seven unrelated stars. Five of them are members of a cluster of stars that is gradually breaking up, and they are travelling around the galaxy together. Since they are about 70 light-years away from Earth, the light we see has taken an entire human lifetime to travel through space to our eyes. The two stars not moving with the group are the last one in the handle and the one in the bowl closest to Polaris. Eventually, they will drift away from the others, but it will take many thousands of years. Our great-great-grandchildren will see exactly the same Big Dipper that we do.

Polaris is the North Star. Somehow, people have come to believe that it is the brightest star in the sky, or at least one of the brightest. The illustration on facing page clearly shows that it is not even the brightest in this section of the sky. Among stars visible from Canada and the United States, Polaris ranks thirty-sixth in brightness. By comparison, Vega is third and Deneb fifteenth. Polaris is of special importance, though, because it is almost exactly above the Earth's North Pole. Anyone facing Polaris is looking due north.

A giant among stars, Polaris is roughly 50 times bigger than our sun and outshines it by 6,000 times. It appears as an ordinary star in the night sky, however, because it is 600 light-years from Earth. We will encounter even bigger and more powerful stars during our tour of the skies in the following pages. For now, though, identifying Deneb and Vega near overhead is the key to advance to the next chart.

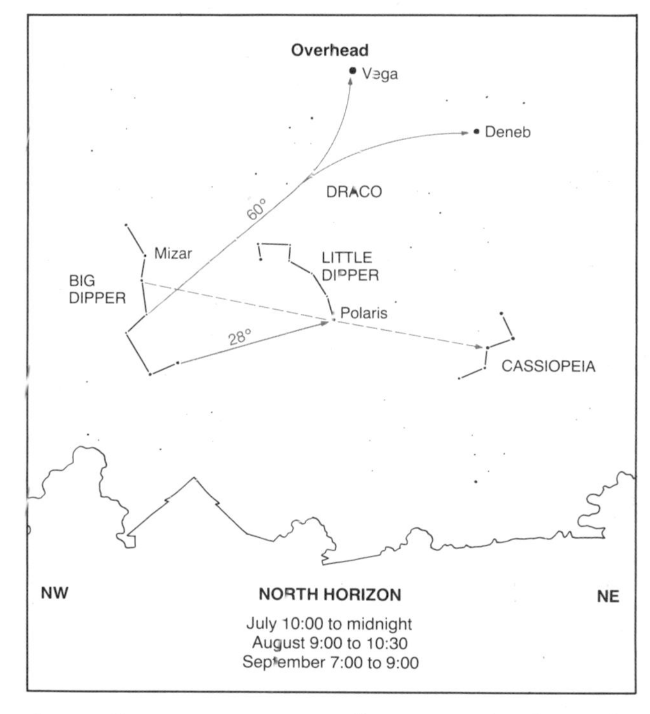

The Big Dipper is the crucial key to the summer sky. Use the Dipper to identify Deneb and Vega in the overhead zone, then advance to the next chart. Cassiopeia unlocks the autumn sky, so become familiar with that constellation as well. The Little Dipper, despite its well-known name, is dim and unimportant apart from its famous member, Polaris, the North Star. The star Mizar has a faint nearby companion star, Alcor. Binoculars show it, though many people can see Alcor without optical aid.

THE SUMMER TRIANGLE: VEGA, DENEB AND ALTAIR

Summer is vacation time for most people, and that means trips to the country, where the night sky is dark and rich with stars. The Milky Way, which looks like a faint trail of white powder on the black backdrop of the night sky, is at its brightest in summer. Part of it is shown in the illustration on facing page.

The Milky Way is seen best on nights with no moon, well away from city lights. Binoculars transform the hazy glow of the Milky Way into a glittering river of stars. Clumps of stars that are mere smudges to the unaided eye become stellar jewel boxes when viewed with binoculars. It is a memorable sight, yet few people who own binoculars ever turn them to the Milky Way. When you do, you will be gazing deep into our star city through the thickest part of our wheel-shaped galaxy.

Three bright stars of summer – Vega, Deneb and Altair – form a large triangle in the southeast section of the sky. The Summer Triangle is bigger than most stargazers expect it to be. It is larger than your hand with your fingers fully spread held at arm's length. Vega, the brightest of the trio, is a nearby star 26 light-years from Earth. It shines 55 times brighter than the sun, is about three times its diameter and has a dazzling white surface much hotter than our sun's.

Just over one degree to the left of Vega is a twin-star system easily distinguished with binoculars. Other stars near Vega form the small constellation Lyra the harp. About 15 degrees to Vega's left is a larger constellation, Cygnus the swan. Cygnus more closely resembles the Christian cross than a swan and is also known as the Northern Cross.

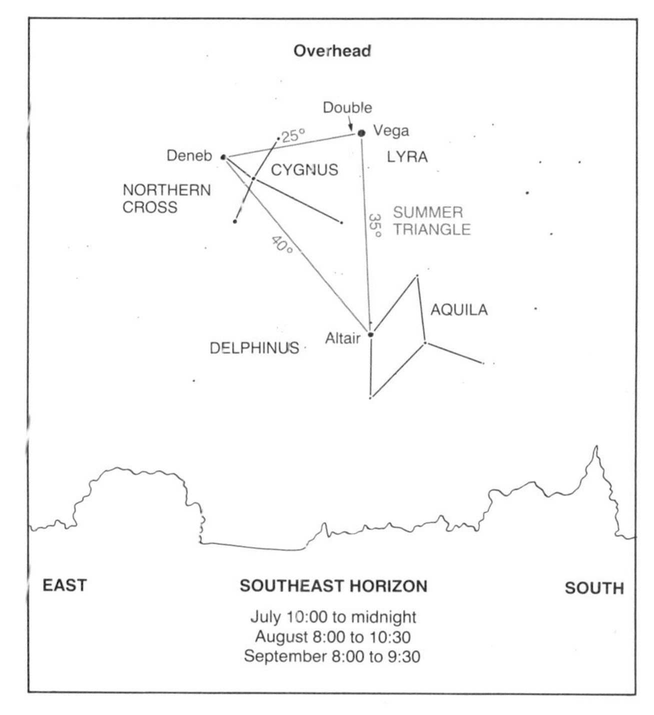

Deneb, the bright star at the top of the cross, is one of the most powerful stars in the galaxy. It would take 60,000 stars like our sun to equal Deneb's brightness. This hot blue-white star is 1,600 light-years from Earth. Altair, the third star in the triangle, is 17 light-years away and 10 times brighter than the sun.

The key to the summer sky is the Summer Triangle, formed by the bright stars Vega, Deneb and Altair. Vega and Deneb are near overhead for most of the summer. They are also seen in the overhead region of the chart on page 53, which shows how they can be located using the Big Dipper. The constellation Cygnus is also known as the Northern Cross. Delphinus the dolphin is a small but easily identified constellation. The Milky Way is brighter in this region of the sky.

CASSIOPEIA, PEGASUS AND THE STARS OF AUTUMN

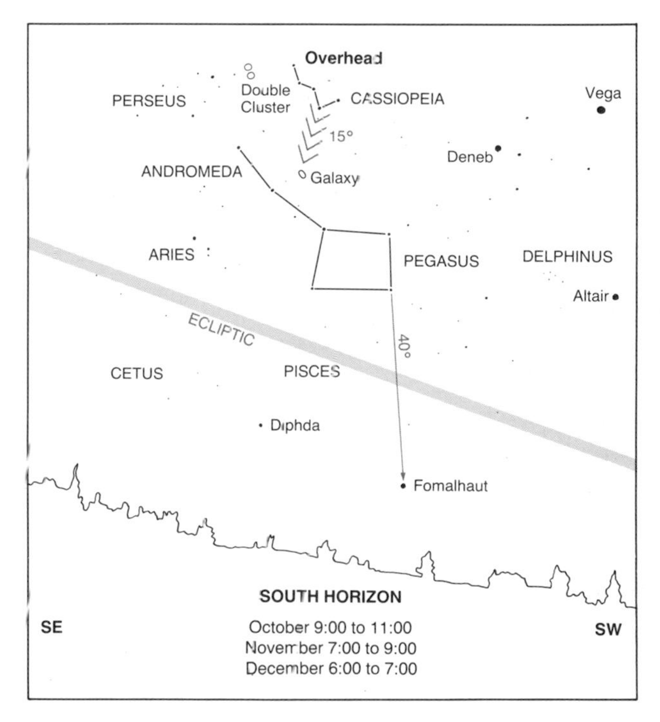

Autumn constellations are the dimmest of the four seasons – they have few bright stars like those that glitter in star patterns seen during the rest of the year. But that is not the whole story. Fall skies display some amazing sights. To find them, you must first identify the W-shaped constellation Cassiopeia, the mythological queen, which appears almost straight overhead throughout autumn.

Four stars in a large square below Cassiopeia represent the constellation Pegasus. Like all autumn constellations, Pegasus does not look anything like the winged horse it was named for. Extending from the upper left corner of Pegasus is the constellation Andromeda, in which the Andromeda Galaxy is found. In the illustration on facing page, it appears as a hazy patch. That's exactly what it looks like in the sky.

The Andromeda Galaxy can be located by using the arrowhead shape of the right side of Cassiopeia like a pointer, as shown on this page. When you look at the Andromeda Galaxy, you are viewing the most distant object visible to the unaided eye. Light from the stars in that galaxy takes more than two million years to reach Earth.

Binoculars make it much easier to see the Andromeda Galaxy and reveal its oval shape. To the left of Cassiopeia is a twin cluster of stars in our galaxy known as the Double Cluster. Like the Andromeda Galaxy, it can be seen as a dim smudge with the naked eye. Binoculars greatly improve the view by showing some of the individual stars in this pair of clusters. Since both the Andromeda Galaxy and the Double Cluster are near overhead, you should look for them while lying in a reclining lawn chair or flat on the ground.

Jupiter will be below Pegasus in 1987 and near Aries in 1988. Saturn will not be in the region covered by the chart until 1993, when it will be to the right of Fomalhaut. From 1994 to 1996, it will be below Pegasus. Mars will be near the centre of the ecliptic track in 1988 and to the left in 1990.

The stars of autumn are dimmer than those of other seasons, so the constellations are not as prominent. The W-shape of Cassiopeia, overhead, is the best starting point—its stars and those in the square of Pegasus are about the same brightness as the Big Dipper's stars. Pegasus is attached to Andromeda. Both constellations are south of Cassiopeia. Another way to identify autumn constellations is to begin with the Summer Triangle and work left through Delphinus to Pegasus.

THE BIG DIPPER AND THE WINTER SKY

The Big Dipper is the first star pattern that amateur astronomers learn to identify because it is unmistakable. There is nothing else like it in the night sky. The Little Dipper is small and inconspicous by comparison. From the city, most of the Little Dipper's stars will be lost in the sky glow. But that's no problem. Polaris, the only important star in the constellation, is as bright as the Big Dipper stars.

Capella is by far the brightest star seen when we face north in winter. The stars in the top of the bowl point directly at it. Be sure to run the sight line two full Dipper-lengths to reach Capella. Actually, Capella can be seen no matter which direction you face because it is almost straight overhead in midwinter. But starting with the Big Dipper will ensure that you find Capella, especially in other seasons when the star is not necessarily overhead. Remember, Earth is always moving – both turning on its axis and travelling in its orbit around the sun – and these motions cause the whole sky to shift slowly, which is why the charts have time limits. These are not rigid limits though. Most charts in this section are useful up to an hour outside the guidelines.

The constellation Cassiopeia, shaped like the letter W, is exactly the same distance to the left of Polaris as the Big Dipper is to its right. Cassiopeia is one of the three most important star patterns in the sky, as we saw on page 48. Above Cassiopeia is Perseus, a straggly group of medium-bright stars that is often passed over because its stars are not arranged in a memorable pattern. However, it is worth looking for Perseus because a beautiful celestial grouping called the Double Cluster lies between it and Cassiopeia.

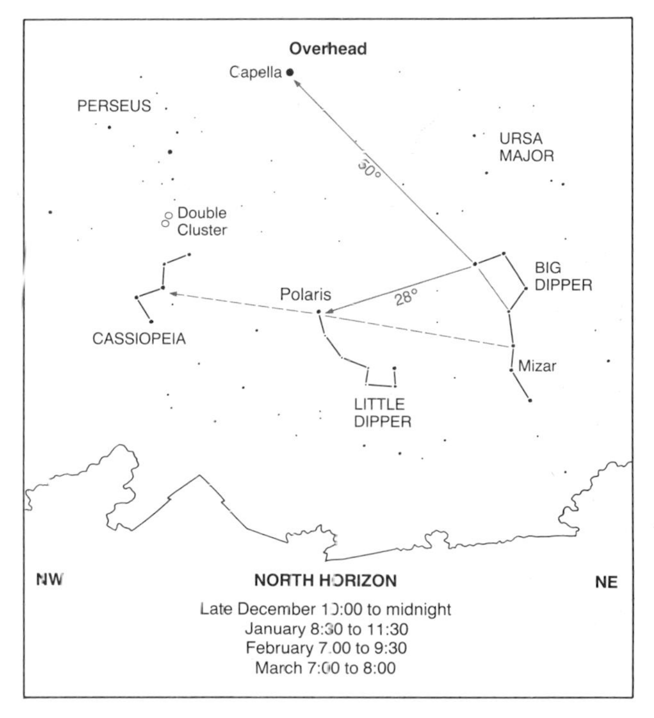

The Double Cluster can be seen with the naked eye in a really dark sky. It looks like a bright knot in the Milky Way. Binoculars improve the view enormously. The Double Cluster is two clumps of stars 7,000 light-years from Earth, in the next spiral arm outward from the arm containing our sun.

The Big Dipper stands on its handle all winter long and guides stargazers to key star patterns. The two stars in the Dipper's bowl opposite the handle are five degrees apart. A sight line extended five times the distance between them, out the top of the bowl, leads to Polaris—the brightest star in the Little Dipper. A line between the two stars in the top of the Big Dipper's bowl points two full Dipper-lengths to Capella, the brightest star in the overhead region during winter months.

ORION AND THE STARS OF WINTER

The stars are more brilliant in winter than at other times of the year – not because the winter air is clearer but because the stars themselves are brighter, on average, than those seen in other seasons. The winter sky show starts with Orion, king of the constellations. Orion has more bright members than any other star grouping. It even looks like the mighty hunter it was named for. Three stars form his belt, two mark his shoulders and two more his knees. Some of the interesting stars in Orion are described on page 63.

Just below Orion's belt is a fuzzy-looking "star." This hazy patch is the Orion Nebula, a cloud of gas and dust 1,500 light-years from Earth that is a nursery for newborn stars. Portions of the cloud are collapsing to form stars. If you look at the Orion Nebula through binoculars, you will see one or two of the new stars buried within it. The nebula is about 25 light-years wide.

Taurus the bull is a zodiac constellation, so planets sometimes appear in it and in nearby Gemini. Aldebaran is the bull's eye, which is somewhat bloodshot since Aldebaran is pale orange. It is a red giant star, 45 times the diameter of our sun. If Aldebaran were to replace the sun, its outer edge would almost reach Mercury's orbit.

Two star clusters – the Hyades and the Pleiades – are contained within Taurus. The Hyades, 140 light-years away, is the nearest distinct cluster of stars to our solar system. Binoculars show it very well, just to the right of Aldebaran. Although three times farther away, the Pleiades cluster is more impressive, since its stars are bunched closer together. Because of its little-dipper shape, the Pleiades is often mistaken for the real Little Dipper, which is actually less prominent.

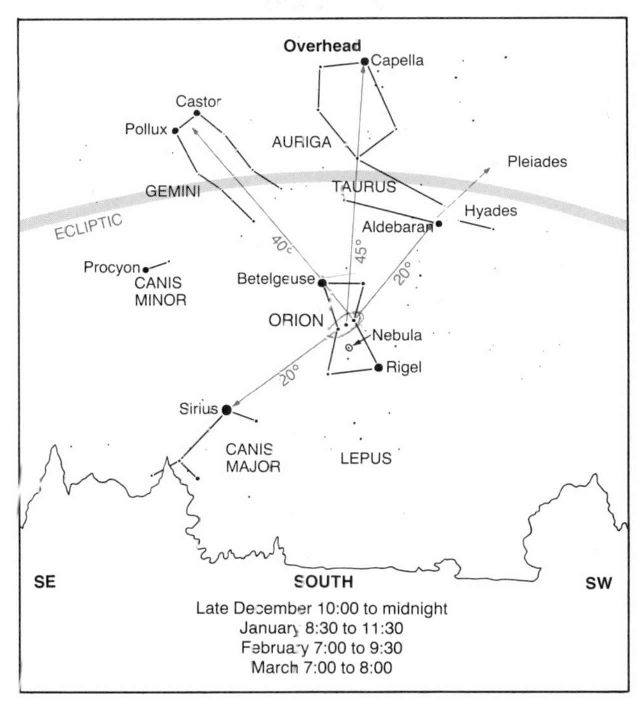

Jupiter will be close to the ecliptic line shown on the chart in 1989, 1990 and 1991. Mars will be visible near the ecliptic from late 1990 to early 1991 and from late 1992 to early 1993. Saturn will not visit this sector before the year 2000.

Orion, the key to the winter sky, is the most important stellar grouping after the Big Dipper. Orion's three-star belt points down to Sirius, the brightest star in the night sky, and up to Aldebaran and the beautiful Pleiades cluster. Orion's other stars are guides to Auriga and Gemini, two prominent winter constellations. Above Orion—almost straight overhead—is brilliant Capella, which is also shown on the chart on page 59. There, you can see how the Big Dipper points to Capella.

STARGAZING FROM THE CITY: ORION STILL IN VIEW

Most people live in or close to a city. The night sky over a city is flooded with light from street lamps, houses, shopping centres, parking lots and other illuminated areas. This makes the sky bright, washing out the faint stars. To see the difference, compare the city view on facing page with the country sky on page 60. Although fewer stars are visible, the bright ones remain. Not only is stargazing possible in the city, but an urban area is sometimes a good place to start because the main constellations are usually all that can be seen.

The winter sky has the brightest stars, so its constellations are the most prominent in city skies. Orion, with its distinctive three-star belt, stands out well. Betelgeuse, Orion's right shoulder, is a red giant and one of the largest stars known. If it were in our sun's place, Earth would be inside it. Its distance from us is uncertain but is probably about 600 light-years. Betelgeuse actually appears red, although not red like a stoplight. It is pale red or orangish. The colour is due to its relatively low surface temperature.

All of the other bright stars in Orion are at the opposite extreme on the stellar temperature scale – bluish white. The brightest, Rigel, is slightly more brilliant than Betelgeuse. Rigel is a real scorcher. It is 50,000 times as powerful as the sun and 30 times its diameter. Rigel is 1,000 light-years from Earth.

Sirius is the night sky's brightest star because it is close – only 8.6 light-years from Earth. It is a hot blue-white star 23 times as luminous as the sun. Sirius is sometimes called the Dog Star because it is the brightest star in the constellation Canis Major, Orion's big dog.

Jupiter will be close to the ecliptic line shown on the chart in 1989, 1990 and 1991. Mars will be visible near the ecliptic from late 1990 to early 1991 and from late 1992 to early 1993. Saturn will not visit this sector before the year 2000.

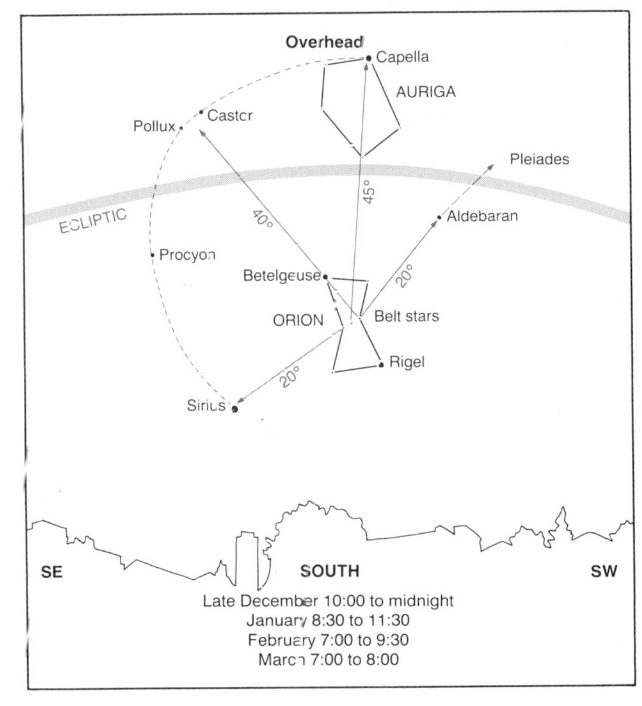

Many of the winter stars are so bright that they are not washed out by city lights. Orion the hunter is especially easy to identify. His three-star belt is unmistakable. The belt serves as a guide to Sirius, the brightest star in the night sky, and to Aldebaran in the constellation Taurus the bull. Orion's stars also provide a starting point for locating the stars Castor, Pollux and Capella. A huge arc of bright stars extending from Capella, near overhead, to Sirius is shown by the dotted line.

LEO AND THE DIPPER OVERHEAD: SIGNS OF SPRING

Early-spring skies are decorated with a beautiful array of bright stars and constellations. The guide to all this is the Big Dipper, high overhead. The locater arrows shown on this page emerge from the Dipper and provide sight lines to the whole sky. To add some comfort to your stargazing, use a reclining lawn chair and a dim flashlight for reading the charts.

Castor and Pollux, the bright pair of stars in the constellation Gemini, mark the imaginary border between winter and spring stars – stars to their right are most easily observed in winter months. Pollux is 35 light-years from Earth, Castor 45 light-years. Castor is actually a six-star system, even though it looks like a single star to the unaided eye and through binoculars.

The premier constellation of spring is Leo the lion. Its brightest star, Regulus, is easily located along the sight line from the Dipper's bowl. Regulus, a blue-white star 160 times brighter than our sun, is 85 light-years from Earth. Leo, unlike most constellations, resembles what it was named for: the lion is reclining with its legs curled under it. The rear legs and haunches are marked by the triangle to the left of Regulus; the head and mane are the stars above Regulus that look like a backwards question mark, with Regulus as the dot.

Between Regulus and Pollux is a misty patch called the Beehive (marked "Cluster" on chart). If you look at it through binoculars, you will understand why: it is a swarm of faint stars – a star cluster 500 light-years away. The Beehive is at the centre of the dim constellation Cancer, which is well known because, like Leo, it is one of the 12 zodiac constellations. These constellations lie on the ecliptic, the pathway of the sun, moon and planets.

Jupiter will appear to the right of Leo in 1991, close to Leo in 1992 and to the left in 1993. Mars will be seen near Leo in 1995 and to the left in 1997 and 1999. Saturn will not be in this part of the ecliptic for the remainder of the 20th century.

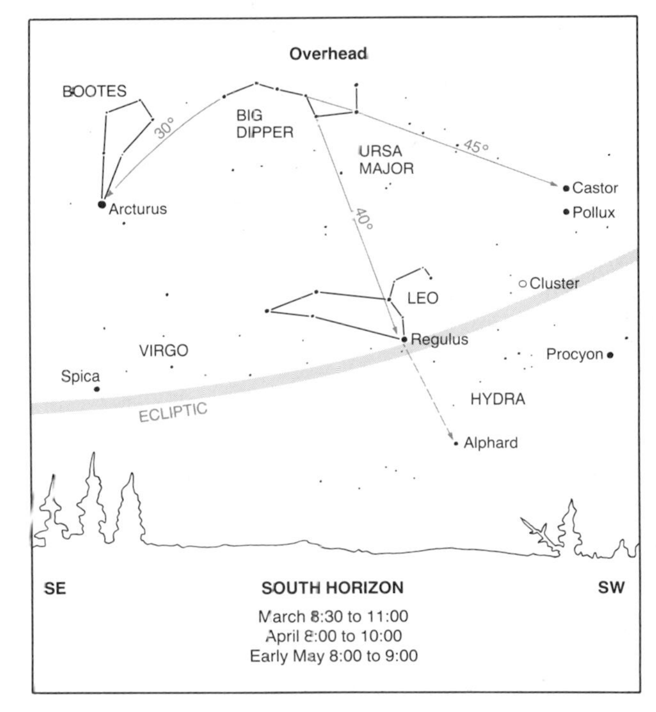

The Big Dipper, almost straight overhead, is the key to the spring sky. Locater arrows (sight lines) from the stars in the Dipper's bowl point the way to Castor and Pollux, Regulus and Arcturus. Once these bright stars have been identified, the constellation patterns are easy to find. Leo is particularly pretty and looks a bit like a lion. The stars on the right side of the chart are also shown in the winter sky chart on page 61. The diagram on the next page adds detail to the left side of this chart.

ARCTURUS, SPICA AND THE SPRING SKY

"Follow the arc to Arcturus, and speed on to Spica." Thousands of astronomers and navigators have repeated this saying to find Arcturus and Spica, the two dominant stars in the spring sky. The arc is the curve of the Big Dipper's handle, which points to Arcturus one Dipper-length away; Spica is another Dipper-length farther on.

The second brightest star visible from Canada and the northern half of the United States, Arcturus is a distinct yellow colour with a touch of pale orange. It is not quite as hot as the sun. (From hottest to coolest, the star colours are: blue, white, yellow, orange and red. The sun is light yellow.) However, Arcturus makes up for its cooler temperature by having a larger surface than our sun – it is 23 times the sun's diameter and 115 times its brightness. At a distance of 36 light-years from Earth, it is the nearest giant star.

Arcturus and the stars directly above it form the constellation Bootes, the herdsman. Once again, the constellation looks nothing like its name; but beside it is a neat semicircle of stars with an apt name: Corona Borealis, the northern crown.

Spica, the leading star in the constellation Virgo, the maiden, is a blue star 220 light-years from Earth that is 1,300 times the sun's brightness. Every star we encounter in our tour of the night skies seems to be much brighter than our sun. Of the 100 brightest stars in the sky, not one is dimmer than the sun. Yet thousands of less luminous stars are out there. In fact, they outnumber the stars brighter than the sun by 20 to 1. But the reason they are not well known to stargazers is that hardly any of them are visible to the unaided eye. It is the luminous giant stars that steal the show, making the night sky the beautiful sight it is.

Jupiter will shine near Leo in 1992 and will move through Virgo in 1993 and 1994. Mars will be seen near Leo in 1995 and to the left of Leo in 1997 and 1999. Saturn will not pass through this area before the 21st century.

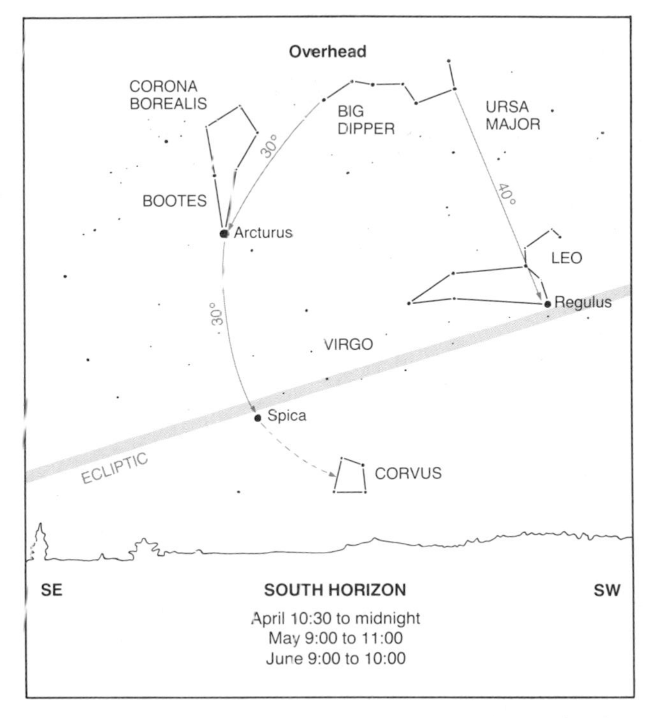

The guide to spring skies is to "follow the arc to Arcturus." The arc in this case is the curve of the Big Dipper's handle, which, if extended one Dipper-length, points to Arcturus, the brightest star in the spring sky. The arc can be continued one more Dipper-length to Spica. Although the Big Dipper is shown nearly overhead, it will appear a little north of overhead from most places south of Montreal and Seattle. A dim flashlight and reclining lawn chair add ease and comfort to stargazing.

BINOCULARS AND TELESCOPES FOR ASTRONOMY

In the chart section on the previous pages, binoculars are recommended for close-up views of the Milky Way, several star clusters, a nebula and a galaxy. People sometimes forget that binoculars are actually two small telescopes. Because they permit viewing with both eyes at the same time, binoculars are easier to use than telescopes. They have another advantage: more of the sky can be seen in one view with binoculars than with telescopes, which operate at higher magnifications.

Binoculars are the best way to get to know the sky after the main constellations have been identified. Experienced amateur astronomers use binoculars as well as telescopes to explore the night sky. Binoculars also reveal the moon's craters, the satellites of Jupiter and the planets Uranus and Neptune. They are better than telescopes for viewing bright star clusters like the Hyades and Pleiades.

Binoculars usually magnify between 7 and 10 times. Numbers engraved on them state the magnification, or power: 7 x 35s magnify 7 times and have main lenses 35 millimetres (1.4 inches) in diameter; 10 x 50s magnify 10 times and have 50-millimetre (2-inch) lenses. Any good-quality binoculars work well for backyard astronomy, but more than 10-power binoculars are not recommended because it is too hard to hold them steady.

But aren't telescopes better than binoculars? Certainly a quality telescope will show much more detail. However, an inferior telescope is not much better than binoculars and is often far more difficult to use. Unfortunately, poor-quality telescopes are sold everywhere – department stores, camera shops and stores that sell eyeglasses.

Good-quality telescopes never have the power prominently displayed on the package or in advertising because high power is not important in astronomy. For example, it takes only 40 power to distinguish the rings around Saturn. Far more important than power are good optics and a solid heavy-duty mounting and tripod. If the telescope feels jiggly, it won't be much use for astronomy. Good optics and rigid mountings are seldom found on telescopes that cost less than $300. An excellent beginner's telescope is a 75mm (3-inch) refractor or a 100mm (4-inch) reflector. Before buying a telescope, read a few practical astronomy books that have sections on selecting and using telescopes. Then find a store specializing in telescopes so you can compare brands, prices and quality. A good telescope will last a lifetime and will provide thousands of memorable views of the universe.

One of the best ways to see what a telescope can do is to attend public "star nights" organized by astronomy clubs. Telescopes are set up in a park or conservation area, and everyone is invited. Not only are the celestial sights on view, but you can talk to people who have a deep interest in astronomy. Watch your local newspaper for announcements of such events.

The author and a young astronomer prepare for a night's observing session with two large telescopes. Few people have sophisticated equipment like this, but very inexpensive telescopes are almost useless for backyard astronomy. A decent telescope costs at least $300. Beginners will have an easier time if they learn their way around the night sky with binoculars before purchasing a telescope. Binoculars are simple to use and show large sections of the sky at one time.

CREDITS

INTRODUCTION: p. 4 NASA.

A COSMIC VOYAGE: p. 5 © Royal Observatory, Edinburgh; p. 6 Illustration by John Bianchi; p. 7 Illustration by Adolf Schaller; p. 8 Illustration by John Bianchi; p. 9 NASA; p. 10 Illustration by John Bianchi; p. 11 Illustration © David Egge; p. 12 Illustration by John Bianchi; p. 13 Illustration by Michael Carroll; p. 14 Illustration by John Bianchi; p. 15 Illustration © David Egge; p. 16 Illustration by John Bianchi; p. 17 Smithsonian Astrophysical Observatory; p. 18. Illustration by John Bianchi; p. 19 National Optical Astronomy Observatories; p. 20 Illustration by John Bianchi; p. 21 Fred Espenak; p. 22 Illustration by John Bianchi; p. 23 Anglo-Australian Telescope Board; p. 24 Illustration by John Bianchi; p. 25 © Anglo-Australian Telescope Board.

ALIEN VISTAS: p. 26 NASA; p. 27 Illustration by Victor Costanzo; p. 28 (top) Illustration © David Egge; p. 28 (bottom) NASA; p. 29 Illustration © Don Davis; p. 30 Courtesy Lowell Observatory; p. 31 © Don Davis; p. 32 NASA; p. 33 Illustration by Adolf Schaller; p. 34 NASA; p. 35 Illustration © David Egge; p. 36 Illustration © David Egge; p. 37 Illustration by Michael Carroll; p. 38 Illustration by Adolf Schaller; p. 39 Illustration by Victor Costanzo; p. 40 Illustration by Adolf Schaller © Edmund Scientific; p. 41 © Anglo-Australian Telescope Board; p. 42 Illustration © David Egge; p. 43 Illustration by Adolf Schaller; p. 45 (top) Illustration © David Egge; p. 45 (bottom) Hale Observatories; p. 46 Illustration © David Egge; p. 47 Illustration by Adolf Schaller © Edmund Scientific; p. 48 Illustration by John Bianchi.

STARGAZING: p. 52 Illustration by John Bianchi; p. 54 Illustration by John Bianchi; p. 56 Illustration by John Bianchi; p. 58 Illustration by John Bianchi; p. 60 Illustration by John Bianchi; p. 62 Illustration by John Bianchi; p. 64 Illustration by John Bianchi; p. 66 Illustration by John Bianchi; p. 68 Wayne Storms.

The companion book to *Exploring the Night Sky* is *Exploring the Sky by Day* by Terence Dickinson (Camden House Publishing; 1988), a look at clouds weather, rainbows and solar haloes as well as astronomical subjects such as moon phases and auroras.

NightWatch by Terence Dickinson (Camden House Publishing; revised edition; 1989) has dozens of easy-to-use charts for stargazing as well as several chapters on how to select and use binoculars and telescopes.

The Universe and Beyond by Terence Dickinson (Camden House Publishing; revised edition; 1992) contains more than 100 full-colour illustrations and photographs of planets, moons, stars, galaxies and quasars. This thorough survey of the universe includes chapters on the search for extraterrestrial life, black holes, the lives of stars and the fate of the universe.

An inexpensive book with good charts of the night sky is *The Monthly Sky Guide* by Ian Ridpath and Wil Tirion (Cambridge; New York; 1987). Ian Ridpath's *Star Tales* (Universe Books; New York; 1988) explains the Greek myths associated with the constellations.

Sky Calendar is an excellent night-by-night illustrated summary of celestial events observable with the unaided eye or binoculars. For current subscription price, write Abrams Planetarium, Michigan State University, East Lansing, MI 48824.

The Nova Space Explorer's Guide by Richard Maurer (Clarkson N. Potter; New York; 1985) is one of the best space/astronomy books of recent years. Also recommended is *Your Future in Space* (Crown; New York; 1986), a beautifully illustrated book about space exploration and the U.S. Space Camp for teenagers in Huntsville, Alabama.

Two books filled with astronomy projects for school or personal enjoyment are *Science Projects in Astronomy* (1990; published by the author; Bill Wickett, 1584 Caudor Street, Encinitas, CA 92024) and *Seeing the Sky* by Fred Schaaf (Wiley; New York; 1990).

For a free catalogue that includes astronomy slide sets, videos, posters and computer software, contact any of the following: The Planetary Society, 65 North Catalina Avenue, Pasadena, CA 91106; Astronomical Society of the Pacific, 390 Ashton Avenue, San Francisco, CA 94112; Hansen Planetarium Publications, 1098 S. 200 West, Salt Lake City, UT 84101; Sky Publishing Corp., Box 9111, Belmont, MA 02178.